CROSS-BORDER INTEGRATION OF RENEWABLE ENERGY SYSTEMS

Cross-border energy trade and integration of renewable energy have become significant for countries and regions to meet demands, minimize costs, and foster socio-economic and climate stability in the dynamic and unstable energy market. This book explores different models of global energy trade between regions and their benefits and challenges with a special focus on India's Northeast region.

Countries in South and Southeast Asia are endowed with abundant renewable energy resources. This book examines the energy mix of the countries such as India, Myanmar, Thailand, Bangladesh, and Bhutan among others and their efforts to achieve more integrated markets and renewable energy integration in the region. It highlights the potential of Northeast India given its rich natural resources and strategic location to harness the potential cross-border energy trade with ASEAN countries. The volume provides analytical perspectives on drivers, constraints, opportunities and barriers, as well as measures that countries could take to address institutional, financial, policy, and governance issues to minimize the total costs of energy security and maximize the social-economic benefits for people in these regions. It identifies the necessary conditions – grid flexibility, policy, market, and regulatory solutions for clean energy trade – and contributes to growth of low-carbon development as well as policy making by focusing on renewable energy integration across borders.

This volume will be of interest to students and researchers of energy and climate studies, environmental politics, trade, and economics and international relations.

This book is freely available as a downloadable Open Access PDF at www.taylorfrancis.com under a Creative Commons (CC-BY-NC-ND) 4.0 license.

Venkatachalam Anbumozhi is the Director of Strategy and Innovations at the Economic Research Institute for ASEAN and East Asia (ERIA), Jakarta, Indonesia.

Bhupendra Kumar Singh is currently Principal, Energy Security at Confederation of Indian Industries (CII), New Delhi, India.

CROSS-BORDER INTEGRATION OF RENEWABLE ENERGY SYSTEMS

Experiences, Impacts, and Drivers

*Edited by Venkatachalam Anbumozhi
and Bhupendra Kumar Singh*

ERIA
Economic Research Institute
for ASEAN and East Asia

Routledge
Taylor & Francis Group
LONDON AND NEW YORK

Designed cover image: Getty Images

First published 2024
by Routledge
4 Park Square, Milton Park, Abingdon, Oxon OX14 4RN

and by Routledge
605 Third Avenue, New York, NY 10158

Routledge is an imprint of the Taylor & Francis Group, an informa business

British Library Cataloguing-in-Publication Data
A catalogue record for this book is available from the British Library

ISBN: 978-1-032-52786-4 (hbk)
ISBN: 978-1-032-56251-3 (pbk)
ISBN: 978-1-003-43316-3 (ebk)

DOI: 10.4324/9781003433163

Typeset in Sabon
by Deanta Global Publishing Services, Chennai, India

CONTENTS

FIGURES

Map

TABLES

CONTRIBUTORS

Editors

Venkatachalam Anbumozhi is the Director of Strategy and Innovations at the Economic Research Institute for ASEAN and East Asia (ERIA), Jakarta, Indonesia. He has more than 20 years of broad experience in the analysis, development, and implementation of climate change, low-carbon growth, green financing programs, and technical assistance projects at strategic and operational levels. Following multi-year assignments in the Japan International Cooperation Agency (JICA) projects in the late 1990s and teaching climate economics, market-based instruments, and international finance cooperation at the University of Tokyo, Japan for several years, Anbumozhi joined Asian Development Bank (ADB) Institute, Japan, in 2008. At ADBI, he assessed the effectiveness of low-carbon green growth policies and developed a technology roadmap for governments to upscale climate change mitigation and adaptation program with public-private investments. As a task manager for the joint ADB-ADBI study, he organized high-level policy dialogues on the innovative financing mechanism for the low carbon energy transition in seven ADB DMCs. He was the coordinator of ADBI-World Bank e-learning courses on micro-finance, wherein South Asia and Sub-Sahara's good practices are shared.

Bhupendra Kumar Singh is currently Director and Head, Energy Security at Confederation of Indian Industries (CII), New Delhi, India, after serving for a short stint at the Ministry of External Affairs, Government of India. A PhD on India's Energy Security from Jawaharlal Nehru University, New Delhi, he has also done an Executive MBA in Oil and Gas from University of Petroleum and Energy Studies, Dehradun. He has written many articles in reputed peer reviewed Journals and News Papers. His article "South Asia Energy Security: Challenges and Opportunities" published in *Energy Policy*, Vol 63 (2013), Elsevier, Washington DC has been widely acknowledged for its landmark recommendations. He had the opportunity to address the Energy Parliamentary Committee on Energy Storage, United Kingdom. He has been a CII member of "ETP Committee", constituted by International Energy Agency (IEA), Paris, France (2010–2011), India representative for South Asia Transmission Utility Regional Network (SATURN) (2012–2013), Member of International Association of Energy Economics (IAEE), USA (2009–2010), Project Director for the Foreign and Commonwealth Office, Government of UK (2015–2016), Project Director for the Department of International Development, Government of UK, Project Director for the Asia Foundation, San Francisco, USA, 2015–2018. He was instrumental in facilitating India–Nepal Power Trade Agreement in 2014. He is constantly working for the enhancement of the cross-border power trade and made a landmark effort to modify the Guideline on Cross-Border Electricity Trade as more transparent, more business friendly and based on the principles of the Open Market. For his exemplary work on cross-border power trade in South Asia, he was awarded with Nepal's Prestigious Award IPPAN Distinguished Person Award in 2021.

Contributors

Daniel del Barrio Alvarez is an assistant professor at the International Project Laboratory in the Department of Civil Engineering at The University of Tokyo, Japan. His research interests include sustainability transition, energy studies, and social innovation. Currently, he conducts studies on the energy transition in Southeast Asia, with a special focus on the role of power connectivity and the promotion of the integration of renewables in the region.

Prabir De is a professor at the Research and Information System for Developing Countries (RIS), New Delhi, India. He is also the Coordinator of the ASEAN–India Centre (AIC) at RIS. He works in the field of international economics and has research interests in international trade and development. He has contributed several research papers in international journals and written books on trade and development.

Kaliappa Kalirajan is an Emeritus professor in the Crawford School of Public Policy at the Australian National University. Research interests: modeling and analysis of sources of growth; regional economic groupings and international trade; micro econometric modeling and policy analysis; and low carbon energy systems in Asia. He has authored, co-authored, or edited 20 books published by reputed international publishers. He has published 170 research papers from 1970 to 2022 in the referred reputed academic journals.

Sunil Malla is an independent research consultant. Prior to this, he worked as a senior researcher and faculty (adjunct) at the Asian Institute of Technology (AIT), Thailand. In the past, he has also worked as a researcher and a consultant for UNEP and Asian Development Bank. He has published several articles in international journals that focus on energy, environmental, and economic issues of Nepal and other developing countries. He received his PhD in economics from the University of Hawai'i–Manoa, USA, ME in energy technology from AIT, Thailand, and BE in electrical engineering from the National Institute of Technology–Rourkela, India.

Adil Khan Miankhel is an applied quantitative economist and has 23 years of public service experience at the policy-making level in many countries in the Australasian region. He has completed several empirical research projects, particularly in the areas of trade, macroeconomics, and energy economics. He was leading the Trade Office in Central Asia during his diplomatic assignment during 2013–2016. He has published widely both individually and jointly in refereed journals.

Kakali Mukhopadhyay is a professor at the Gokhale Institute, Pune, India and an adjunct professor for the Agricultural Economics Program at McGill University, Canada. She has been a senior advisor of the E3-India Model, a collaborative initiative of RAP, Vermont, USA, and Cambridge Econometrics, UK, providing high-level inputs for its development and validation. She has received a number of international fellowships from the World Bank, the Asian Development Bank, the Indo-Dutch Program and the Ford Foundation. Her research focuses on energy and environment, trade, air pollution and health, REI, R&D, GVC, economics of health and Nutrition, and food safety.

Shankaran Nambiar is a senior research fellow at the Malaysian Institute of Economic Research, Kuala Lumpur, Malaysia. He has served as a consultant for many Malaysian ministries and international organizations abroad. Dr Nambiar has published widely and actively comments on economic and geo-economic issues for various news and media outlets such as Nikkei Asian Review and Jakarta Post. He holds a PhD in Economics from Universiti Sains Malaysia.

Vishnu S. Prabhu is a PhD scholar at the Gokhale Institute of Politics and Economics, Pune, India.

Ehsan Rasoulinezhad is an associate professor of economics and deputy dean at the Faculty of World Studies of Tehran University, Iran.

Priyam Sengupta is a PhD scholar at the Gokhale Institute of Politics and Economics, Pune, India. He is an independent researcher with much experience in doing economic research and analysis using the Input–Output framework.

Citra Endah Nur Setyawati is the Research Associate in Energy Affairs at the Economic Research Institute for ASEAN and East Asia (ERIA), Jakarta, Indonesia. She is involved in a number of studies related to the subject of energy.

Farhad Taghizadeh-Hesary is an associate professor of economics at the School of Global Studies of Tokai University, Japan, and the vice president of The International Society for Energy Transition Studies (ISETS) based in Australia.

Govinda Timilsina is a senior economist at the Development Research Group (Research Department) of the World Bank, Washington DC, US. He has 25 years of experience across a broad range of development economics (e.g., infrastructure, energy, urban transportation) and environmental economics, particularly climate change. Prior to joining the World Bank, Dr. Timilsina was the Senior Research Director at the Canadian Energy Research Institute, Calgary, Canada. Earlier, he was with the Asian Institute of Technology, Bangkok, Thailand. In Nepal, he worked with the Nepal Telecom and Nepal Electricity Authority in the early 1990s.

PREFACE

As the global community seeks to ramp down electricity sector emissions, cross-border energy trade and integration of renewables into it are key to building a cost-effective, low-carbon resilient power system. Cross-border grid interconnections could relieve burdens related to excess power generation capacity and help in an efficient evacuation and distribution of renewable energy across the regions. European experiences highlight that advancing intercontinental electricity transmission systems could reduce up to 10% of carbon dioxide emissions alone. In addition to environmental and poverty alleviation and developmental impacts, the cross-border energy trade could reduce energy prices, mitigate supply shortages, and power shocks, incentivize further market integration, manage regional and subregional resource endowment differences, and facilitate sustainable development goals.

Previous studies by ERIA demonstrate existing benefits of renewable energy trade across Lao PDR–Thailand–Malaysia and India–Nepal including displacement of coal generation and much greater integration of offshore wind and solar photovoltaic investments into the existing system and improved electric grid efficiency.

However, in most of the world, the potentials and multiple benefits are mostly untapped. Integrating renewable energy resources into global or regional power grids is challenging due to variability, uncertainty, and flexibility of energy supply, financial support, physical infrastructure, and regulatory mechanisms. Grid interconnections across borders could contribute to a higher share of renewable electricity in participating Member States. The necessary low-carbon energy transformation and energy security are required to necessitate significant cooperation and commitment between governments, businesses, and consumers.

The eight states in India's North Eastern Region (NER) and their neighbors in Southeast Asia are experiencing significant economic growth and forging close ties in the energy sector. But progress in renewable energy integration requires more comprehensive strategies to reduce the gaps in transmission infrastructure, investment, and trade facilitation measures.

ERIA has started examining the best practices needed in tackling barriers to developing new renewable energy-based grid infrastructure and the institutions that bedevil cross-border energy trade projects. This book is based on papers presented and discussed for the ERIA research project on "Integrating Border Zone NER–India with Neighboring Economies: Identifying Energy Markets, International Economic Linkages and Capacity Building for Sub-Regional Cooperation".

This project brought together leading international energy economists and global best practices experiences and generated common insights and understanding. The chapters highlighted several ways to overcome barriers to benefit cross-border renewable energy projects and trade. The most basic requirement for the integration of renewables into cross-border energy trade is a grid integration – a hardware. Yet building new cross-border transmission involves navigating permitting and siting processes across multiple jurisdictions. Power trade also requires countries to work together on "soft infrastructure" – institutions and markets that can facilitate exchange. How do countries allocate costs of hard infrastructure upgrades? Australia and America's highly integrated "soft infrastructure" and flexible markets enable higher penetration of renewable energy cross-border trade regions.

The eight chapters offer reviews and reflections by academic leaders and practitioners covering a range of issues. It recounts the lessons learned and explores the way forward for NER–India. This book captures the collective wisdom of the contributors. We are extremely grateful to all the contributors for their support to this book project and for providing constructive criticism to improve the quality of the book contents. We also thank the anonymous reviewers for their comments and suggestions. In preparing the manuscripts for publication we had excellent assistance from Ayu Pratiwi Muyasyaroh and Abigail Gracia Balthazar and we acknowledge their contributions.

This book is being published as part of ERIA's effort to produce knowledge products that can be used to promote energy security and sustainable development, one of the three priority themes of research. We are confident that this book will contribute to policy development and academic understanding in an area where new insights are urgently needed. We hope this book will also help decision-makers in India and ASEAN to set up and implement robust policy measures and sustainably manage their critical energy resources for the long-term development of their people.

1

CROSS-BORDER ENERGY COOPERATION AND TRADE

Impacts, Challenges and Policy Implications

Venkatachalam Anbumozhi, Bhupendra Kumar Singh, and Citra Endah Nur Setyawati

1.1 Background

Cross-border energy cooperation and trade involves two or more states, and is significant in developing a set of framework conditions in support of energy security. This has been emphasized as a way to meet regional energy demand and supply in an optimal way, provide affordable energy, and contribute to climate stability (Ecofys, 2015; Caldés et al., 2018). Indeed, the economic imperative for cross-border energy connectivity and integration of renewable energy is strong in several parts of the world (Meus et al., 2019). In South East Asia, Kimura and Shi (2014) proved that cross-border energy trade and integration of renewable energy into grid systems would minimize the total costs of energy security and maximize the social-economic benefits. The estimated cost saving could go up to USD 1.3 billion a year in Asia if these strategies are promoted as a part of long-term economic and infrastructure development, fostering innovations for social inclusion (ADB, 2015; Kutani & Li, 2014).

Embarking on cross-border energy trade and integration of renewables, mainly hydro power in South and South East Asia, has brought multiple co-benefits to the member countries as illustrated in the electricity exchange between India and Bhutan as well as Lao PDR and Thailand. In the case of India, Anbumozhi, Kuttani, and Lama (2019) showed through model-based assessments that a reduction in final energy cost is feasible by promoting cross-border energy trade among South Asian countries. They also concluded that sub-regional renewable energy zones could become a powerful source for low-carbon energy transition, if and when the participating

DOI: 10.4324/9781003433163-1

countries liberalize their energy markets in support of more cross-border energy trade.

From a financial investment perspective, expanding and integrating renewable-based grid systems in support of cross-border energy trade will allow infrastructure investors and consumers to harness the advantages of economies of scale, thus enabling them to have better access to larger financial resources and to tap relatively cheaper clean energy sources (Anbumozhi, Kuttani, & Lama, 2019). For example, the joint renewable energy auctions in Europe contributed to low electricity prices paid by the consumers in Norway and Germany, which is part of the European Union. Von Blücher et al. (2019) found that this would also result in decreased public funding for cross-border energy projects.

The climate benefits of cross-border energy trade are due to the flexibility of large-sized energy systems that can integrate higher shares of modern renewable energy sources such as wind, solar, and biomass. This is because, with larger energy supply systems, there is greater resilience in reducing the vulnerabilities underlying changes in electricity demand. For example, solar photovoltaic produces differing energy intensities across any grid system (Sayigh, 2018). Increased integration of them into cross-border energy trade could result in reducing the gaps in peak electricity demand and supply needs. Further, they permit the allocation of energy supply between the grids resulting in overall renewable energy system efficiency.

1.2 Hierarchy Models of Integrated Energy Systems and Cross-Border Energy Trade

Multiple models of integrated energy systems that have demonstrated positive economic, environmental, and social impacts and experiences in renewable energy system integration exist across the world. In principle, there are two models of cross-border energy market integration. The first is on the geographical scale and direction of energy trade, which range from limited unidirectional to full-scale energy market integration. The second is on a time scale ranging from short term to medium to long term.

Table 1.1 shows the examples of cross-border energy market integration and trade models facilitated through investments in grid connectivity. It includes limited cross-border bilateral energy trade to complete and unify the market and operations. Altogether, they represent a hierarchy of regional cooperation and integration of renewable energy into the grid system. Economic and social benefits could be maximized through greater integration of renewables and increased energy trade across borders, but involve complex institutions.

As the models show, this hierarchy of cross-border energy market evolution can further be classified into three sub-categories in terms of energy

TABLE 1.1 Models of cross-border integrated energy systems

Type	*Example Practice*
Bilateral and unidirectional.	Power export from Lao PDR to Thailand.
Bilateral and bidirectional.	India–Bangladesh.
Multilateral, multidirectional among differentiated markets.	Southern African Power Pool.
Multilateral, multidirectional among harmonized markets.	EU energy markets.
Unified energy market with differentiated operations.	Nord Pool.
Integrated energy market and unified operations.	Pennsylvania–New Jersey–Maryland Interconnection.

Source: Authors

trade: bilateral, multilateral, and unified. Under the bilateral integration model, cross-border energy trade occurs between two administrative territories. In some cases, this energy trade may be unidirectional, as in the case of Lao PDR and Thailand in South East Asia, and in some other cases, there may be a necessity of wheeling, where the grid connectivity is established for transit purposes. Sub-regional authorities are involved in that model for designing the flows of power, as evidenced in the South African Power Pool.

Multilateral modes of integration often involve more than two countries that started trade in electricity due to supply and demand gaps. However, there may exist large variations within the trading countries in terms of energy production structure, planning of distribution networks, and incentives for the integration of renewables. In several countries, it was found that establishment institutions for cross-border energy trade facilitate more market integration, though they do not necessarily replace national energy administrations and local market authorities.

An IEA (2019) study on power system integration found that under unified models of energy market integration, there are instances, as in Europe, where new regional institutions could take control of all functions of national and local energy authorities. From the perspective of time horizons, cross-border renewable energy market integration can involve the progressive implementation of pricing reforms over a longer period. On the other hand, power purchase agreements and real-time electricity dispatch could occur on short time horizons. Between these two extremes are policy areas and actions that may be governed by regional cooperation agreements, such as the ASEAN Plan of Action in Energy Cooperation (APAEC), which outlines five work streams in support of a unified regional market. Nevertheless, they need to be supported by establishing a mechanism for short-term energy supply and demand forecasts across the borders and/or information on day-ahead scheduling. As illustrated in Table 1.1, the hierarchy of a completed cross-border energy market integration and energy trade does not guarantee an orderly evolution process, but is hindered by several technical, economic, and financial factors (Kerres, 2020).

1.3 Economic Implications of Cross-Border Energy Trade and Renewable Energy Investments

Several studies have already quantified the benefits of increased integration of renewables in cross-border energy trade initiatives from the perspective of investors and Transmission System Operators (TSO). Actually, many international energy trade initiatives start with a long-term outlook for energy security and progressive implementation of supporting policies such as the establishment of regional day-ahead markets, as happened in Europe. Nevertheless, there exist examples of first establishing short-term markets as practiced in the South African Pool to harness long-term benefits. Meus et al. (2019) have summarized the investor perspectives on the benefits of integrating renewables and cross-border energy trade. In fact, several modes of energy market integration exist, though the benefits vary.

An overview of energy security, economic and environmental sustainability benefits coming from the increased integration of renewables in Organization for Cross-regional Coordination of Transmission (OCCTO) Japan, Operators European Network of Transmission System Operators for Electricity (ENTO-E), and Southern African Power Pool (SAAP) are summarized in Table 1.2.

Primarily, the cross-border energy trade projects serve the goals of secured energy supply, reduced electricity prices, increased consumer choices, and decreased carbon emissions. As could be seen from the case of the European Network of Transmission System Operators for Electricity (ENTSO-E), all are achieved through the development of a network of transmission systems. SAAP and OCCTO also have similar arrangements which have varying economic costs. According to Ecofys (2015), in three years starting from 2006, European TSOs collected about EUR 1.6 billion in congestion rents and wheeling alone. A study by EC (2013) found that integrating the networks of the Netherlands, Germany, and the Nordic regions would reduce costs by around EUR 100–180 million per year. In the USA, it was estimated that the costs of establishing a centralized market for Pennsylvania–New Jersey–Maryland Interconnection, is less than the efficiency gains which would eventually compensate initial investments IEA (2019).

In South and South East Asia there is no single regulatory authority responsible for developing regional cross-border energy trade. As a result, regional and sub-regional planning exercises tend to be done more on a

TABLE 1.2 Co-benefits of increased integration of renewables and cross-border energy trade

OCCTO	**Energy Security:** Securing stable electricity supply. **Economy:** Suppressing electricity rates to the maximum extent possible. **Sustainability:** Expanding choices for consumers and business opportunities.
ENTSO-E	**Energy Security:** Pursuing coordinated, reliable, and secure operations of the interconnected electricity transmission networks; promoting the adequate development of interconnected grid and investments for a reliable, efficient, and sustainable power system. **Economy:** Providing a platform for the market players by proposing and implementing standardized, market integration, and transparency frameworks. **Sustainability:** Facilitating secure integration of new power generation resources, particularly renewable energy.
SAAP	**Energy Security:** Provide a forum development of world-class, robust, safe, efficient, reliable, and stable interconnected energy systems; increase accessibility to rural communities. **Economy:** Facilitate the development of competitive electricity market. Give the end-user a choice of electricity suppliers. **Environmental Sustainability:** Ensure sustainable energy development through sound economic, environmental, and social practices; aim to provide the least cost, affordable clean energy.

Source: Authors based on ERIA (2017)

country basis. An Economic Research Institute for ASEAN and East Asia (ERIA) study (2017a) estimated that the formation of a single, unified market for South East Asia could reduce the total energy system costs by 10%, and potential financial savings would be more than what could be achieved through a set of optimal bilateral agreements. It was complemented by another study, ERIA (2017b), which found that increased coordination of four TSOs within Europe left a saving of EUR 3.3 billion by reducing reserve power costs over a period of five years.

1.4 Challenges in Integration of Renewables and Cross-Border Energy Trade

Though there is an economic case and environmental benefits of renewable energy integration into cross-border energy trade, in several countries development of interconnected markets can be challenging. The IEA (2019) study has identified four phases of renewable energy deployment and its implications on cross-border energy trade.

Phase 1: The impact of modern and variable renewable energy on the hydro power-dominated energy mix is relatively small and does not have a significant impact on the operations of cross-border energy trade and transmission system operations.

Phase 2: The total share of modern variable solar, wind, and biomass energy increases to a level point where the impacts on cross-border energy trade are visible, but not big enough to make any changes in the operations of hydro power-dominated energy systems.

Phase 3: Hydro, modern renewable, and hybrid power change significantly over a short time period and warrant due consideration for increased flexibility in the system operations.

Phase 4: Integration of modern and variable renewable energy power is high enough, which results in new stability problems for the system operations, requiring changes in both technical and operational design.

The above four phases are directly related to the share of variable renewable energy in the energy mix and cross-border energy trade. However, it is important to acknowledge the fact that the four phases have no definite delineations. The impacts of an increasing proportion of renewable energy within the national energy mix system are also specific to domestic and international technological pricing policies rather than the cross-border energy trading system itself (Sayih, 2018). However, moving up from Phase 1 to higher levels is directly related to the increased share of renewables in the energy mix which could increase the cross-border energy trade. It is a critical factor, as it implies energy market integration is a useful tool in increasing the renewable energy supply at the national or sub-regional level. The relative proportion of renewables in large, integrated grid systems may be lower than the absolute share in isolated local systems if some participating states have invested more solar, wind, and biomass power generation than others involved in the energy trade.

Moreover, as in any multilateral trading system of goods and services, some countries will get more benefits than others, and in selected cases, there may be net negative impacts. The key challenge for renewable energy integration across the border is how to allocate investment costs across several economic actors and share the operational benefits evenly with consumers.

Regarding the primary benefits of energy security, there are several technical issues that need to be tackled in a comprehensive way. First, individual states participating in the energy trade tend to have an aspiration that they need to remain self-sufficient in energy production, which means they cannot rely on electricity purchased from neighboring states. Second, the high integration of imported renewable energy with non-renewable energy could further increase the risk of blackouts when the supply is cut short due to technical reasons. Third, synchronized energy systems across the borders must deal with unexpected cross-border transit challenges.

This is a big challenge in states with higher shares of modern variable renewables which tend to fluctuate drastically as climate patterns find shifts. Ultimately, in any interconnected energy system's national pricing policies, subsidy regimes and taxes have implications on cross-border energy trade. For example, renewable energy policies that support local investment in renewable energy generation at subsided rates can result in surplus energy available, and beyond a certain level may result in price distortion if the demand for energy flows are not coordinated. Local energy production capacity and energy consumption patterns can also bring in system inefficiencies.

1.5 Planning for Resource Adequacy

Cross-border energy trade and renewable energy integration start with the development of grid interconnections. As discussed before, the multilateral interconnection may take several years to materialize and will involve multiple stakeholders, including very different governmental institutions, private sector operators, and Independent Power Producers. Further, cross-border energy trade could be complicated, as it necessitates the scaling down of sub-regional level energy development planning and making new bilateral trade pacts to distribute investment costs fairly among the parties. Ideally, any sub-regional planning should look at overall resource adequacy, which requires the aggregation of local renewable energy investment plans into sub-regional grid connectivity plans, which necessitates deepened discussion and agreements on perceived benefits, risks, timeline on implementation, and scenario planning. Regional institutions are also needed and need to be established in time to coordinate the planning in a collaborative way. It is necessary because cost-sharing formulas are often based on the beneficiary pay principles wherein the total costs and benefits are shared in proportion to the calculations and received in return for the benefits. In several cases, it is justifiable to divide the costs along lines that are based on bilateral political agreements than on economic rationale. It may also be a real challenge in sharing the costs when the benefits are real but too diffused to be fully captured in the short and medium term. In that scenario, a specialized regional institution may come forward to facilitate the developments through technical support and financial assistance programs.

After two or more energy markets are integrated through the establishment of several transmission lines, specific arrangements are necessary to guide energy trade between the relevant grids. Adequate provisions are crucial as they allow for optimization of supply and demand of electricity derived from renewable energy sources. Cross-border energy trade involving renewable energy could be made as simple as possible without any financial agreements wherein bilateral or multilateral institutions agree on the quantity of electricity to be traded based on the market prices. Energy market structure within countries decides the price mechanism. For example, in the case of energy trade between Malaysia and Singapore, two different market structures exist. Singapore is fully market-oriented where wholesale retail markets are fully liberalized, while Tenaga Nasional Berhad in Malaysia is a fully government-owned utility company. There are no fundamental barriers for both Malaysia and Singapore to be connected by transmission lines.

Energy market integration across borders and renewable energy planning at the sub-national level are also adopting similar bottom-up approaches, as in the case of the USA, where the Northwest Power and Conservation Council developed a regional plan for power trade among four states – based on utility-level integrated resource plans. Several such resource adequacy models have been documented by The European Topic Centre on Climate Change Mitigation and Energy (ETC-CME) (2020).

1.6 Institutions and Governance Frameworks for Cross-Border Energy Trade

Once a policy decision has been made to maximize renewable power in cross-border trade, it is the governing institutions that will determine how effective it will be in achieving the benefits in time. In fact, the issue of institutions and governance frameworks matters more than any other economic factor because it sets the direction for plans and their implementation.

The primary governance requisite for any cross-border energy cooperation project is the strong political will to foresee the long-term benefits, analyze the risk, reach an agreement on costs, and share the benefits (EC, 2016). Tackling the legal complexities of any international agreement requires effective coordination and wider consultations. That is why the planning of cross-border energy trade needs sufficient lead time, appropriate allocation of human and financial resources for negotiations, and for all key stakeholders to be invited and involved to ensure all specific issues are identified and addressed. In the Lao PDR and Thailand bilateral energy trade negotiations, it took nearly ten years for a pioneering agreement to be reached. In Europe and elsewhere, countries only came to an agreement on cross-border energy trade when the net social, economic, and environmental benefits were estimated to be larger than the investment costs and implementation risks. This shows the importance of recognizing the risks and identifying de-risking mechanisms, as well as quantifying the costs and benefits as much as possible.

Nevertheless, countries participating in the cross-border energy trade should keep a flexible approach in the initial stages and focus on how the plan of cooperation would help them meet the strategic goals of energy security, economic efficiency, and sustainability gains in a more cost-efficient manner (EC, 2013). Both economic costs and socio-environmental benefits could be impacted by macroeconomic policies such as taxes, subsidies, and investment barriers. Nevertheless, sometimes it is difficult to monetize the environmental benefits and social costs which would also be changing over time. While a balanced distribution of total costs and net benefits among participating countries must be pursued, making a perfectly equal distribution would be very difficult, as observed in the case of the ASEAN Power Grid (APG) (ERIA, 2017a).[1]

In reality, APG can be seen as a bottom-up planning approach, wherein no regional institution with regulatory functions is established to oversee the development of transmission lines, market creation, and harmonization of rules, which is considered to be a prerequisite for cross-border energy trade. But the ten Association of South East Asian Nations (ASEAN) member states collectively set aspirational targets for establishing power grid interconnections with their dialogue partners providing support for analytical works in studying the feasibility and demonstrating the benefit of energy market integration. Three ASEAN Interconnection Masterplan Studies have been completed and an ASEAN Power Grid Consultative Committee was formed to help the APG agenda move forward.

Macroeconomic policy frameworks are also important for facilitating cross-border energy trade (IEA, 2019). Thailand, for example, imports hydro power-generated electricity from Lao PDR. The structure of these energy imports, however, differs significantly from other models such as the energy trade between the USA and Canada. The hydro power in Lao PDR was built by Thailand and other countries and is often treated as Independent Power Producers (IPP) often under the control of Thailand's utility company, Energy Generating Authority of Thailand (EGAT). Though Lao PDR generates electricity from the dams located in its territory, it consumes little electricity and has limited ability to influence the power plants' production capacity. This is a similar case with India–Bhutan bilateral energy trade, wherein Bhutan earns revenue from the electricity trade from India but not necessarily electricity. Both Thailand and India's national utility companies are fully owned by their governments, and their actions, when it comes to cross-border energy trade, can be seen as reflecting the countries' geopolitical and economic preferences for international cooperation. From Lao PDR's viewpoint and Bhutan's perspective, these cross-border energy trade arrangements are suboptimal, because they are generating financial assets within their national territory that use their local resources but only derive a partial benefit from them. The governance frameworks are also starting to change. Rather than discussing the imperatives for new hydro plants being built under an IPP model, current debates are more focused on moving to grid-to-grid trading models. Driven by a desire to increase the diversity of its power generation sources, the most recent Power Development Plan (PDP) of Thailand, released in 2015 (EGAT, 2015), includes a planned increase in imports from 10% of its current energy-generating mix to between 15% and 20%.

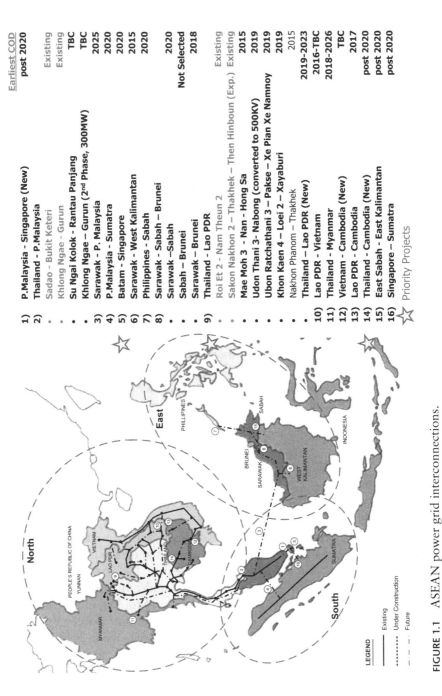

	Earliest COD
1) **P.Malaysia - Singapore (New)**	post 2020
2) **Thailand - P.Malaysia**	
• Sadao - Bukit Keteri	Existing
• Khlong Ngae - Gurun	Existing
• **Su Ngai Kolok - Rantau Panjang**	TBC
• **Khlong Ngae – Gurun (2nd Phase, 300MW)**	TBC
3) **Sarawak – P. Malaysia**	2025
4) **P.Malaysia – Sumatra**	2020
5) **Batam - Singapore**	2020
6) **Sarawak - West Kalimantan**	2015
7) **Philippines - Sabah**	2020
8) **Sarawak – Sabah – Brunei**	
• **Sarawak –Sabah**	2020
• **Sabah – Brunei**	Not Selected
• **Sarawak – Brunei**	2018
9) **Thailand - Lao PDR**	
• Roi Et 2 - Nam Theun 2	Existing
• Sakon Nakhon 2 – Thakhek – Then Hinboun (Exp.)	Existing
• **Mae Moh 3 - Nan - Hong Sa**	2015
• **Udon Thani 3- Nabong (converted to 500KV)**	2019
• **Ubon Ratchathani 3 – Pakse – Xe Pian Xe Namnoy**	2019
• **Khon Kaen 4 – Loei 2 – Xayaburi**	2019
• Nakhon Phanom – Thakhek	2015
• **Thailand – Lao PDR (New)**	2019-2023
10) **Lao PDR - Vietnam**	2016-TBC
11) **Thailand - Myanmar**	2018-2026
12) **Vietnam - Cambodia (New)**	TBC
13) **Lao PDR - Cambodia**	2017
14) **Thailand - Cambodia (New)**	post 2020
15) **East Sabah - East Kalimantan**	post 2020
16) **Singapore – Sumatra**	post 2020
☆ Priority Projects	

FIGURE 1.1 ASEAN power grid interconnections.

Source: ASEAN Center for Energy (2015). ASEAN Plan of Action for Energy Cooperation (APAEC) 2016–2025

In the case of different developmental stages and energy demands, good communication plays a transformational role, ensuring that the benefits of energy cooperation are sufficiently explained to the public (Ecofys, 2015). While the cross-border energy infrastructure development projects enjoyed a wider backing from international financial institutions like Asian Development Bank, the participating countries of Lao PDR and Thailand are committed to transparency and regularly exchanged information on their investment strategies, not only to international investors but also avoided any negative spillovers that could influence power purchase agreements. It is also important to have a highly coordinated approach in discovering new energy pricing policies as in the case of the Mekong sub-region. This alleviated the concerns that cooperation might interfere with the effectiveness or efficiency of domestic policy measures, which is commonly reported as a barrier to cross-border energy cooperation (IEA, 2019).

1.7 Designing Domestic Energy Market Structures for Enhanced Cross-Border Trade

The European Energy Union is an example of how to develop advanced multinational power market frameworks and cross-border energy trade (IEA, 2019). There are also, however, many countries that have differing internal power market structures. Japan, for example, has a common energy policy set at the national level but a fragmented utility structure, as shown in Figure 1.2.

However, quite a large number of energy market reforms are underway in several countries in South East and East Asia, and, when completed, will create an integrated market framework for the entire country (ERIA, 2017a). Japan's retail energy market is already open to investors at the national level, and the Japan Electric Power Exchange (JEPX) is also functioning and facilitating new markets. However, the volume of trade on JEPX has been at low levels because energy trade has been limited to national utilities alone. Shinkawa (2018) reported that the proportion of the share in trading to the total energy demand in the Japanese energy market remains as low as 6.8% in 2017. As domestic energy market reforms move forward, new participants will certainly enter the markets and the utilization of the power exchange will likely increase. From the perspective of cross-border energy market integration, however, the two most relevant markets, a balancing market and a capacity market, are still under development. Notably, both will be organized by OCCTO which has significant responsibilities in promoting increased energy market integration among the regional energy power companies, as well as ensuring the security of supply for the power system in Japan. Improving cooperation across the jurisdictions on both balancing and resource adequacy procurement is fundamental to the goals of these market reforms.

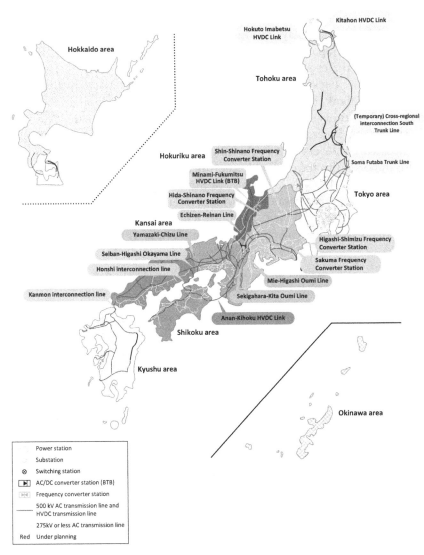

FIGURE 1.2 Cross-regional interconnections in Japan.

Source: ERIA (2017a)

1.8 Energy Market Integration in India and Cross-Border Energy Trade

In the past 30 years, India has developed its domestic power system from a single-buyer model to a fully harmonized system. The five separate grids of the 1990s have been transformed into fully synchronized operational zones within the integrated market over a period of 20 years. There are four key milestones in the chronological evolution of the Indian power market

TABLE 1.3 Energy market integration and grid synchronization in India

Time	Stages of Synchronization
Pre-1991	Five – North, South, East, West, and North East regional grids.
October 1991	East and North East synchronized to form the central grid.
March 2003	West synchronized to the central grid.
August 2006	North synchronized to the central grid.
December 2013	One synchronized grid.

structure as illustrated in Table 1.3, which has been jointly governed by central and state government jurisdictions.

Currently, two power exchanges are operating in India, namely the India Energy Exchange (IEX) and the Power Exchange India Limited (PEX). They compete in the markets of electricity buyers and sellers across the state judications. The East and North East grid systems are often faced with the challenge of system inefficiency and underinvestment.

There is increasing policy interest in strengthening energy connections between the North Eastern Region (NER) of India and major economic centers in South and South East Asia (GOI, 2015). The NER comprises the contiguous eight hilly states, namely Arunachal Pradesh, Assam, Manipur, Meghalaya, Mizoram, Nagaland, Sikkim, and Tripura. This region accounts for 9% of India's geographical area and contributes 3% to the country's gross domestic product (GDP).

However, given its rich natural resource base and strategic geographical location, the NER has a high potential to become a powerhouse in terms of energy, trade, and investment (Anbumozhi et al., 2018). The NER is unique in terms of the energy resource development and economic growth opportunities it offers. About 98% of the region's borders form India's international boundaries with China, Bangladesh, Bhutan, and Myanmar. Given its strategic location, the region can be developed as a base for India's growing economic links not only with the ASEAN but also with its South Asian neighbors.

Over the past several years, India has been a part of several South and South East Asian regional initiatives that aim to deepen its energy market and economic integration. These include the South Asian Free Trade Agreement (SAFTA), the Bay of Bengal Initiative for Multi-Sectoral Technical and Economic Cooperation (BIMSTEC), and the Asia-Pacific Trade Agreement (APTA). Further, as a part of its "Act East" policy and "Indo Pacific Strategy," India has intensified its economic relationship with the members of the ASEAN and countries belonging to the East Asia Summit (EAS). The centrality of the NER is critical in effectively pursuing these initiatives. However, the progress of energy and trade and investments in the eight states of the NER of India has been relatively limited due to regulatory bottlenecks, policy misalignments, and information gaps (ADB, 2015).

India's growing relationship with its neighboring countries including the ASEAN has two major tracks. The first is the development of an enabling environment through the adoption of comprehensive economic partnership agreements with the region as well as with individual countries, and improvement in regulatory procedures. The second track involves the building of energy and transport infrastructure under the fabric of ASEAN–India connectivity, which includes the Trilateral Highway, Kaladan Multi-modal Transit Transport Project, and the Mekong–India Energy and Economic Corridors.

To promote people-to-people linkages and trade in the local regions, India has been setting up Border Haats across the NER sharing borders with neighboring countries. The following Border Haats are already operational: (i) Kalaichar (Meghalaya–Bangladesh border), (ii) Balat (Meghalaya–Bangladesh border), (iii) Kamlasagar (Tripura–Bangladesh border), and (iv) Srinagar (Tripura–Bangladesh border). Similar initiatives have been proposed at the NER–Myanmar border regions. However, these Border Haats are not meant for attracting large-scale investments. To attract investments, particularly foreign investments, in NER borders, broader energy, and economic connectivity, including the establishment of economic zones (BEZ), are necessary. Surrounded by an international border, BEZ is a natural choice for the NER, and may help India become more economically engaged with Myanmar, the gateway to the ASEAN.

Anbumozhi, Kutani, and Lama (2019), through extensive surveys, scenario analysis, and stakeholder consultations, identified the steps that can be taken to realize the full energy trade and investment potentials of the region. They concluded that given its strategic location, the NER can be developed as an energy and economic trading hub for India's growing economic links with South East Asia and other South Asian neighbors. The NER has the potential to grow faster than its current pace, provided the region builds cross-border energy and economic trade and production links, particularly with Myanmar and other South East and East Asian countries such as Thailand, Lao PDR, Cambodia, and Viet Nam. Stronger industrial production networks supported by energy trade would enhance the integration process.

Nevertheless, the barriers to cross-border energy trade are multiple. NER state authorities are unaware of the magnitude of the multiple potential benefits of cross-border energy and the state-level regulatory factors limiting it. To facilitate industrial production networks, the NER would essentially need a strong presence of dynamic enterprises and an uninterrupted flow of energy, particularly renewable energy. The presence of small and medium industries (SMEs) in the NER today is sparse and regional supply chains are relatively weak when compared to ASEAN member states. The NER needs to investigate renewable energy more extensively in the closer economic integration process that India has been seeking with its eastern neighborhood through its "Act East" Policy and "Indo Pacific Strategy." The NER has a lot of wind, solar, and small hydro power potential with a very minimal installed capacity. BEZs would also be developed into smart cities in the NER along international corridors. Some of the future supply chains could include SMEs in processed food, software, electronics, education, health, and garments sectors (GOI, 2015).

1.9 Scope of the Book and the Contents of the Chapters

Set against this backdrop, this book has the following objectives:

a) Assessing the energy security status and creating an energy security index for the North Eastern Region with all forms of energy, including technology and innovation that could happen in small hydro, solar, and wind energy, and its impact on cross-border energy trade.
b) Assessing the stepwise improvement in absorptive capacities of the neighboring countries to electricity exports from the NER and other regions in terms of market, transmission capacities, regulatory framework, investment destinations, and other institutional capacities.
c) Analyzing good international practices in cross-border energy trade in order to identify driving forces, impacts, and experience in facilitating enhanced cross-border energy trade between the NER of – India and the ASEAN.

Maximum exploitation of renewable energy sources such as solar, wind, biomass, and small hydro dams is one of the best ways to facilitate economic growth and achieve energy security as well as promote cross-border energy trade. The chapters in the book conduct a critical analysis of the following issues: (i) key energy security indexes for the seven states and a composite one for the region; (ii) key problems and prospects of cross-border energy trade and renewable energy integration between the NER and South East Asia to find a possible solution to bring consensus among the governments of these regions; (iii) formulating investment plan and strategies in power and renewable energy sector of this region – based on international best practices in cross-border energy trade; and (iv) recommendations to the state governments of the NER on formulating comprehensive strategic policy to cross-border energy trade.

The broad method of analysis is based on indexing and quantitative assessment of parameters for cross-border energy cooperation and trade, as shown in Table 1.4.

TABLE 1.4 Parameters for renewable energy integration, energy security, and cross-border energy trade

Component	Quantitative Assessment	Energy Security and Sustainability Index
Development of domestic resources.	Self-sufficiency.	Total primary energy supply. Self-sufficiency ratio. Reserve production ratio. Reserve consumption ratio.
Acquisition of cross-border resources.	Diversification of import source. Diversification of energy source. Dependence on energy-exporting regions.	Diversity in sources of oil, gas, and coal import countries. Total primary energy supply/electricity.
Securing a resilient supply chain.	Reliability of energy supply. Build cross-border energy infrastructure.	Reserve margin of generation capacity. Power outage frequency/duration. Commercial energy access ratio.
Management of demand.	Investments in energy efficiency.	Total primary energy supply/Gross domestic product ratio.
Preparedness for sudden supply disruptions and shocks.	Build strategic oil and gas reserves.	Days of prolonged oil shocks.
Environmental sustainability and climate change.	Carbon intensity.	Carbon emissions/domestic energy supply ratio. Carbon emissions/Fossil fuel ratio. Carbon emission/GDP ratio.

Source: Authors

1.9 Conclusions and Key Findings

Cross-border infrastructure development and energy trade are key enablers for the integration of higher shares of renewable energy. More wind, solar, and biomass could be integrated into hydro-powered electricity grids to bring resilience to the system operations. An analysis of international experiences, impacts, and drivers presented in the chapters illustrates a number of potential benefits which could be summarized as

(a) Energy security: Integration of renewables into cross-border grid connections enhances regional energy security by enabling effective utilization of locally available renewable energy sources and sharing across the country borders for the common benefit of consumers. The resultant energy reserve capacity is found to ensure a more reliable power supply across the states or regions.

(b) Economic benefits: Cross-border energy interconnectors and trade could contribute to considerable energy-saving gains during times of peak capacity needs. Economic benefits could be maximized by allowing cost-effective renewable energy generating units within the cross-border zones and enabling power trade between two or more grid system operators.

(c) Environmental benefits: Integration of renewables across the borders reduce the cost of carbon abatement. Cross-border interconnectors enable flexibility in low-carbon renewable energy resources to be shared across different jurisdictions. A larger geographical area for cross-border energy trade will also enhance the resilience, the supply and demand variability, shocks, and uncertainties, of renewable energy generation and thus reduced carbon emissions.

The energy demands of the ten member states of the ASEAN and India have grown by 55% in the past 15 years (IEA, 2022). The abundant renewable energy sources are not distributed evenly across the ten countries of the region and the NER. Enhanced energy trade between India's NER and the ASEAN can play an important role in climate-smart development by enhancing energy security through fuel diversification and reducing the import dependency on fuels. The cost of renewable energy, especially solar PV and onshore wind, has been decreasing significantly in the last decade, and soon could become as cost-competitive with conventional generation technologies such as oil and gas. The ASEAN Plan of Action for Energy Cooperation (APAEC) 2016–2025 set a renewable energy target of 23% energy supply by 2025 in the region. Cross-border interconnection and energy trade will be a key enabler for the integration of higher shares of renewable energy.

Optimizing cross-border flows through multilateral trade from ASEAN to India based on the existing cross-border infrastructure available in the NER can also reduce annual operational costs of the power sector in Myanmar and Thailand. Multilateral trade with India could enable further avoidance of coal power plants. This emphasizes the importance of maximizing the share of renewable energy. With a high share of renewable energy, cross-border energy trade helps to reduce the impact of natural and built environments. The challenge of renewable energy curtailment with increased share can be addressed with increased cross-border trade capacity between ASEAN (say Myanmar) and India (NER).

India's NER is an emerging powerhouse. In terms of hydropower, the seven states of the NER have the potential of about 58,971 MW which is almost 40% of the country's total hydro potential in 2017. Additionally, the region also has an abundant resource of coal (1,630 million tons) and natural gas (195.68 BCM) for thermal power generation (GOI, 2015). If all of its energy potentials are harnessed, the North Eastern Region can become an energy hub and may facilitate cross-border power trading being geopolitically located between South East Asia and South Asia. In view of the energy resources complementarities, the creation of a sub-regional power market for renewables and encouragement of competition among energy producers, both public and private, will enhance business opportunities. It could also be conceived as a long-term goal for efficient clean energy delivery of services to the consumers in India and the ASEAN.

A harmonious regulatory regime, dedicated network of transmission lines, uniform grid code, appropriate electricity tariff fixation, continuous investment flow, rigorous energy diplomacy, and strong political support are the determinants of higher renewable energy integration and enhanced cross-border energy trade, as we can see from the chapters of this book. Nevertheless, participating governments should adopt the following value-added comprehensive strategies as a priority to enhance regional energy security and thereby increase investment, business opportunities, and ultimately the economic well-being of the people.

1.10 Policy Implications

Strategic actions for renewable energy (RE) integration in the NER are as follows: enhanced strengthening of coordination of scheduling and dispatch with neighboring states for better access of least cost generation is one of the prerequisites of RE integration in the North East Region of India, and state-of-the-art automated load and RE forecasting systems should be adopted urgently. Scheduling and dispatch could be upgraded to 5 minutes from the current 15-minute basis. The creation of model Power Purchase Agreements (PPAs) for RE that move away from must-run status and employ alternative approaches to limit financial risks should be encouraged. Integration issues at the distribution grid, including rooftop PV and utility-scale wind and solar that are connected to low voltage lines, must be addressed. RE generators should provide grid services such as automatic generation control and operational data. Central Electricity Regulatory Commission/State Electricity Regulatory Commissions (CERC/SERC) should issue regulations to enable policy-related interventions. Central Transmission Utility/State Transmission Utilities (CTU/STU) should upgrade the technologies and make necessary investments to handle the intermittency through appropriate technical interventions. Availability of substantial flexible generation that can ramp up very quickly should be ensured. Availability of storage devices as a reserve should be increased. The procurement mechanism of flexible generation in assuring grid stability should be strengthened. Fast trading of power at power exchange to manage variable generation should be promoted. Fair price discovery and compensation of flexible resource providers should be chalked out. Balancing areas to reduce variability by offering more balancing resources/demand should be expanded. Evacuation of power through Green Energy Corridors from the regions having high concentrations of renewable energy sources should be sharpened with advanced technologies. This also requires the upgradation of grid operational protocols to ensure that renewable energy does not affect the grid.

Development of border infrastructure: a prerequisite for energy availability is the existence of requisite infrastructure to facilitate transmission/transportation of energy and goods supplies. Lack of infrastructure has had an adverse impact on the exploitation of resources and the ability to access available resources. Building a robust infrastructure may promote regional trade which could be undertaken through interconnection points between neighboring member states.

The border trade infrastructure at the NER is still inadequate to support the rising trade volume. In other words, the NER needs drastic improvement in border infrastructure. The benefits of connectivity corridors will flow only when border infrastructure is upgraded to facilitate trade and investment at the border region.

Creation of state-of-the-art Border Economic Zones: India may consider building state-of-the-art BEZs across the India–Myanmar region. We should facilitate industries in the border region. The BEZs may promote local industries with scale and border benefits. Like Greater Mekong Subregion (GMS) economic corridors, several cross-border connectivity corridors are under construction such as the Trilateral Highway, Kaladan Multimodal Transit Transport Project, etc. A possible extension of the India–Myanmar–Thailand Trilateral Highway to Cambodia, Lao PDR, and Viet Nam are also under consideration. Since the state-of-the-art Border Economic Zones need huge investment and efficient management, Public–Private Participation (PPP) should be encouraged.

Promotion of the market to play a greater role: market force is the main driver of the electricity trade. Let the private sector of a country deal with the private sector of other countries to develop joint ventures for electricity generation, cross-border transmission, and electricity trade. Hierarchical interactions at the company level that provide energy market forecasting in order to reduce the magnitude of risk arising from uncertainty should operate simultaneously with country- or government-level interactions, such as mutually agreed upon market preferences and regulatory choices. The electricity trade should not be a government-to-government deal, rather it should be a deal between the private sectors. The governments could facilitate the deals, develop rules and regulations to facilitate the trade, and strictly enforce the regulations.

Promotion of electricity trade as basic commodity trade: electricity is a commodity, a basic need. It should not be treated as a strategic good for political maneuvering. Politicians could try to use power generation and transmission projects and cross-border trading arrangements to maximize their political mileage (e.g., winning the next election). It does not help to encourage cross-border electricity trade. Relaxing the central/federal government controls and empowering the power utilities and traders to make decisions on the trade deals is a critical factor for the successful expansion of cross-border or regional electricity trade. Countries in South Asia and South East Asia should also give more freedom to their state or provincial governments or electricity utilities regarding their electricity business. If, for example, Indian states have more authority to trade electricity with neighboring countries, there might be a higher level of cross-border electricity interconnections and trade.

Setting up regional institutions: the creation of regional institutions for cross-border electricity integration is needed to develop common rules and regulations to facilitate cross-border electricity interconnection and trade. These institutes include regional electricity regulatory commissions, regional advisory commissions, regional leadership forums, etc.

Demonstrate leadership: leadership and business vision are critical. It is leadership that facilitates cross-border electricity cooperation. Joint leadership between the trading partner countries or even electricity utilities or private sector parties to facilitate cross-border electricity cooperation and trade is necessary.

Reform the electricity markets to facilitate cross-border electricity trade: power sector reforms significantly facilitated the cross-border electricity trade because the reforms allow the electricity traders to quickly grasp the market opportunities in a very short interval of time, such as with day-ahead markets. The fully liberalized market is efficient as it helps to balance supply and demand at the lowest costs. The market reforms harmonize the electricity markets through consistent rules and regulations across the power markets between the borders. The lack of market-oriented reforms in the South East Asian region has restricted either the entry or establishment of supporting institutions. The presence of supporting institutions such as power exchanges, traders, and private sector participation can lead to improvement in domestic power sector performance and can advance progress toward regional integration. Financial reform of loss-making distribution companies (DISCOMs) is also very important in order to ensure the smooth flow of electricity from the point of generation to the point of consumption.

Knowledge-based decision and stakeholders' engagement: generation and utilization of knowledge and engagement of stakeholders are critical for informed decision-making regarding the cross-border electricity interconnection. Collaborative decision-making processes are important to build trust and reduce public opposition. Therefore, the development of any new cross-border electricity system must facilitate public involvement, such as citizens, civil society, and relevant stakeholder groups. Explaining the wider benefits of the proposed project or trading arrangements to all involved communities at the local, national, and regional levels can increase the public acceptance of those initiatives. Making grants available for studies in the early stages of a project development phase, when the cost–benefit analysis and the technical feasibility of the project are still not clear, is important.

Persistence efforts until the mission is accomplished: since cross-border electricity cooperation and trade involve multiple governments, the path is not straightforward. The process of cooperation could be lengthy. It could face resistance from the participating countries because cross-border transmission lines pass through several political jurisdictions, including those that may not benefit directly from the transmission lines and electricity trade. This would not only cause delays and cancellation of the existing projects but also signal risks to future projects. Potential investors would get discouraged. Cross-border electricity interconnection and trade take time. It took more than 100 years in both Europe and North America. However, once the process starts, it moves forward if there is political will.

Strong political will: countries in the region are at different stages of evolution in terms of power market design. There is minimal political commitment to liberalize the sector by undertaking market-oriented reforms in the region. Deeper levels of integration will require that national power markets be at similar stages of reform to address concerns regarding the benefits of integration. India is the only country in the region that has progressively implemented power sector reforms to an extent that the new amendments, which have been proposed but are yet to be approved, address further segregating the wires and supply businesses in the distribution sector. Although other countries in the region have undertaken power sector reforms, these measures were mostly on the institutional side and had a limited impact on the design of the power market.

Integrated strategy for the South East Asian nations: it is essential to conduct an overall assessment, optimization, and adjustment of planned cross-border power connectivity plans to provide detailed information for public and private decision-makers about the quantity, quality, and location of APG and GMS master plan projects, technical standards, and institutional capacities.

It also needs the development of a comprehensive renewable energy investment roadmap as a strategy to show bold leadership in removing the barriers to integration and to make new investments more cost-effective at the grid level through regulations, incentives, and capacity building for taking credit risks.

Earmarking of financial resources for power market integration by expanding the ASEAN Infrastructure Fund to drive private investments with clear policy signals should be encouraged. India should play a proactive role in promoting the NER as a catalyst in connecting with the ASEAN Power Grid as well as in long-term projects such as the "One Sun, One World, One Grid" (OSOWOG).

Formation of common business-friendly legal and regulatory framework: energy markets in individual member states are governed by individual legal, regulatory, and policy frameworks. Such a divergence in the mandate of regulators across the region can create a problem in coordination. Therefore, as a first step, these regulators need to work together to develop a roadmap for harmonizing the relevant regulations.

The legal, policy, and regulatory risks multiply in the case of a cross-border transaction as the trade arrangements need to deal with multiple frameworks. The risk is further exaggerated by the fact that the investments required in such projects are substantial and the investors need to be given an assurance of return on investment. To mitigate these risks, the South and South East Asian region needs to move toward a common legal and regulatory framework to govern the cross-border transactions.

Creation of an enriched energy database: a comprehensive and reliable energy database for the region would facilitate better estimation of business, trade, and cooperation benefits intra- and inter-region. Moreover, sharing of information is a strong confidence-building measure that could pave the way for better cooperation and trade within the region. The database would also aid in better planning at the regional and sub-regional levels.

The database may be built with respect to electricity demand and supply patterns of the participating countries along with the generation profile and the prevailing tariff regimes. Such a database would also aid in developing operational guidelines for the selected trade option. This will help to generate confidence among the cooperating partners and will facilitate them to evaluate the benefits of cooperation from their respective perspectives. An existing ample reserve of technical knowledge and expertise available within the region can be mobilized to provide support to those who need them.

Promotion of private sector participation: a very important source of funding in the region is funded through increased private sector participation and Public–Private Partnerships. At present, most of the energy sector is operated by publicly owned utilities and these have not had a good financial performance for a variety of reasons. As a result, the investment capability of these utilities is limited. Therefore, thrust has been given to encourage private sector participation in regional trade, either individually or jointly with public utilities.

An advantage that private sector participation has is that these entities are seen as neutral parties driven by commercial principles as compared to state-owned utilities. This adds to the credibility of their involvement in cross-border trade. Also, it is perceived that the involvement of the private sector in regional trade would speed up the implementation of projects. However, member state governments need to provide policy, as well as regulatory and contractual clarity so as to increase investments in the region. They need to move toward a more transparent policy and regulatory mechanism and also develop clear contracts with well-defined provisions related to taxation and royalties.

Harmonization of tariff: commercial barriers in terms of tariffs limit the ability to identify and evaluate the real delivered cost of power. The basic problem emanates from the fact that power trading is most often treated in political rather than in commercial considerations. Governments need to provide enabling agreements covering the sector and the deal should be left to commercial decision-making specifically to ensure its long-term sustainability and to inject transparency and accountability. These tariffs need to be rationalized based on a pricing policy that is sound and sustainable. Harmonization of pricing policy among all participating countries is essential to create an environment conducive to sustainable energy cooperation. The mechanism of sharing of interconnection costs needs to be worked out, etc.

To ensure sound commercial and operational efficiency of the utilities: utilities are the ultimate bulk buyers of the traded energy, and it is very essential that they become financially viable and capable of honoring commercial contracts. To ensure sustainable energy trade it is essential that the government focuses on the commercial and operational efficiency of the utilities. Metering, billing, and collection efficiencies of South and South East Asian utilities are below the required standards, which erode their financial soundness and creditworthiness for trade and investment. Smart energy systems such as smart metering, microgrids, and nano grids should be promoted in the NER to improve the efficiency levels and reduce leakages. Furthermore, since RE is variable in nature, investment in energy storage technologies should be complementary to integrating RE sources in the regional grid as well as for expanding cross-border trade.

Harmonization of political mindset: national policies and political mindset proved to be major inhibitors of energy cooperation, despite its potential to bring about improvements in terms of greater energy security and efficiency to the participating countries.

Under the prevailing ideology of economic self-sufficiency, power trade is not a priority issue in the countries of the region. Power trade should not be seen as just simple trading in power, but needs to be understood as a policy that has the potential to bring about necessary change in the quality, reliability, and efficiency of power supply and thereby accelerate the process of commercial and economic growth.

A step-by-step approach of cooperation can be adopted to gradually build up the confidence of the participating countries and utilities. Apart from this, the harmonization of macro policies in the energy sector among participating countries is also essential.

Integrated demand–supply management: it is observed that in South and South East Asia energy in general and electricity are being utilized in an inefficient manner. This is evident from one of the highest energy consumptions per dollar GDP of countries in the region as compared to other countries of the world. Apart from this, the transmission and distribution losses of electricity in these countries are some of the highest in the world. Another aspect of low efficiency is attributed to an imbalance between demand and supply which can be significantly improved through various demand-management measures including power trading among the countries of the region.

As demand for electricity increases, management proceeds from being supply-oriented to being resource-oriented and then to being demand-oriented. In the early stages, measures to increase supply are taken to satisfy the demands as they develop. As demands increase, various energy resources are utilized until these prove to be constraints. Finally, the demand management measures which are relatively more complex are adopted.

However, the need is for an integrated demand–supply management that attempts to identify and implement initiatives that improve the use of energy supply capacity by altering the characteristics of the demand for energy. This comprises several tools involving a mix of pricing, other load management, and conservation strategies designed to increase the incentives for a more efficient use of energy.

Since variable renewable energy (VRE) is dependent on climatic conditions for power generation, an energy hierarchy should be determined, such that various energy options integrated with the grid are available to meet the increase in demand across various time periods. As India and the NER still have a long way to go in achieving 100% transition toward cleaner energy sources for power generation, an energy hierarchy by prioritizing RE sources followed by fossil fuel sources will assist in ensuring grid reliability, flexibility, and stability. Furthermore, India has achieved a significant milestone of "One Nation, One Grid, One Frequency." Synchronization of regional grids has enabled the transfer of electricity from power surplus regions to power deficit regions between the NER and the rest of India as well. In order to improve the efficiency of power trade, the upgradation of transmission capacity in the NER is also necessary.

Considering the environmental impact of power sector reforms: India has committed to work toward the Sustainable Development Goal of ensuring sustainable, affordable, reliable, and modern energy for all (SDG 7), as well as tackling the adverse impacts of climate change (SDG 13). In this regard, promoting RE sources such as solar, wind, and hydro not only for regional energy consumption but also for cross-border electricity trade will ensure the incorporation of cleaner energy sources not only in the NER and India but in our neighboring countries as well.

As could be seen from the chapters in this volume, several common developmental, environmental, and social narratives exist for enhancing cross-border trade between the NER of India and the ASEAN, with Myanmar being the gateway. Integrated development of abundant hydro, solar, and wind potential is sufficient not only for meeting domestic demand but also for exporting to other countries with an expanded regional power grid system. Full potentials of cross-border energy trade with an increased share of renewable energy can be harnessed if the planning process is coordinated and operational practices are harmonized across state jurisdictions with an aim to provide an affordable, reliable, and clean energy supply.

Note

1 ASEAN Center for Energy. (2015). *ASEAN Plan of Action for Energy Cooperation (APAEC) 2016–2025*. Page 18. https://aseanenergy.org/2016-2025 -asean-plan-of-action-for-energy-cooperation-apaec/.

References

ADB. (2015). *Cross-border power trading in South Asia: A techno economic rationale*. Manila: Asian Development Bank.

Anbumozhi, V., Kutani, I., & Lama, M. K. (2019). Energizing connectivity between Northeast India and its Neighbours. Economic Research Institute for ASEAN and East Asia, Jakarta. https://www.eria.org/publications/energising-connectivity -between-northeast-india-and-its-neighbours/.

Anbumozhi, V., Velautham, S., Rakhma, T. F., & Suryadi, B. (2018). Clean energy transition for fueling economic integration in ASEAN. In Bhattacharya (Ed.), pp. 331–347. *Handbook of energy in Asia*. London: Routledge.

Caldés, N., Del Río, P., Lechón, Y., & Gerbeti, A., et al. (2018). Renewable energy cooperation in Europe: What next? Drivers and barriers to the use of cooperation mechanisms. *MDPI Open Access Journal, 12*(1). *Energies 2019, 12*(1), 70, 1–22. https://doi.org/10.3390/en12010070.

EC. (2013). Commission staff working document - Guidance on the use of renewable energy cooperation mechanism, SWD(2013), 440 final.

EC. (2016). Commission staff working document - Impact assessment accompanying the document proposal for a directive of the European parliament and of the council on the promotion of the use of energy from renewable sources (recast), SWD/2016/0418 final.

Ecofys. (2015). Driving regional cooperation forward in the 2030 renewable energy framework. Ecofys Netherlands B.V.

EGAT. (2015). Thailand power development plan 2015–2036. Energy Policy and Planning Office, Planning Institution. Retrieved from http://www.egat.co.th/en/ images/about-egat/PDP2015_Eng.pdf.

ERIA. (2017a). Study on the formation of the ASEAN power grid generation and transmission system. Economic Research Institute for ASEAN and East Asia, Jakarta.

ERIA. (2017b). Study on the formation of the ASEAN power grid transmission system operator institution. Economic Research Institute for ASEAN and East Asia, Jakarta.

ETC/CME. (2020). Cross-border regional cooperation for deployment of renewable energy sources. European Topic Centre on Climate Change Mitigation and Energy. https://www.eionet.europa.eu/etcs/etc-cme/products/etc-cme-reports/etc-cme-report-6-2020-cross-border-regional-cooperation-for-deployment-of-renewable-energy-sources.

GOI. (2015). India's strategy for economic integrtaion with CLMV, Ministry of Commerce and Industry, Government of India.

IEA. (2019). Integration of power systems across the borders, International Energy Agency, Paris. https://www.iea.org/reports/integrating-power-systems-across-borders.

IEA. (2022). South East Asia's energy perspectives, International Energy Agency, Paris.

Kerres, P. (2020). Renewables cross-border cooperation in the energy community. Policy Brief. European Union. Retrieved from https://euneighbourseast.eu/news-and-stories/publications/policy-brief-renewables-cross-border-cooperation-in-the-energy-community/.

Kutani, I., & Li, Y. (2014). Investing in power grid inter connection in East Asia. Economic Research Institute for ASEAN and East Asia. Retrieved from https://www.eria.org/publications/investing-in-power-grid-interconnection-in-east-asia/.

Meus, J., Bergh, K., Delarue, E., & Proost, S. (2019). On international renewable cooperation mechanisms: The impact of national RES-E support schemes. *Energy Economics*, *81*, 859–873.

Sayigh, A. (2018). Transition towards 100% renewable energy. Selected Papers from the World Renewable Energy Congress, Springer.

Shinkawa, T. (2018). Electricity system and market in Japan. Ministry of Economy, Trade and Industry. Retrieved from www.emsc.meti.go.jp/english/info/public/pdf/180122.pdf.

Von Blücher, F., Gephart, M., Wigand, F., Anatolitis, V., Winkler, J., Held, A., … Kitzing, L. (2019). Design options for cross-border auctions. Design and Technology University, AURES. https://backend.orbit.dtu.dk/ws/portalfiles/portal/200641748/AURES_II_D6_1_final.pdf.

2

INTEGRATIVE STRATEGY FOR PROMOTING APPROPRIATE RENEWABLE ENERGY RESOURCES, ENERGY SECURITY AND CROSS-BORDER ENERGY TRADE IN THE NORTH EASTERN REGION OF INDIA

Bhupendra Kumar Singh

Energy security is an important ingredient of national security. For the importing country, it emanates from needing continuous access of energy supply to sustain economic and commercial activities, but for the exporting country, it has to maintain continuous access of the energy market to sell their energy products. For the consumer, it is the continuous access of energy supply at affordable prices. However, the concern of climate change is one of the biggest challenges of energy security that the world is facing today. Renewable energy has the potential to play a critical role in addressing the threat of climate change. It is for this reason that harnessing renewable energy at the optimum level is an important agenda for every country today.

Though India has surplus power, the lion's share of that power is thermal. That is why the main thrust of the Government of India is to promote renewable energy with a goal of increasing its installed capacity in 2022 to 175 GW. Hydro power, solar and wind energy are important renewable sources of energy which can help create a sustainable energy security ecosystem. In this, the North Eastern Region (NER) can play an important role as it has abundant hydro power and solar power. The geographical location of North East India also promotes cross-border power trade with South and South East Asia which is basically renewable in nature.

2.1 Energy Security of North East India

The NER has the potential to become a powerhouse in terms of energy, trade, and investment because of its abundant natural resources and strategic location on the borders with Nepal, Bhutan, Bangladesh, and Myanmar.

DOI: 10.4324/9781003433163-2

Natural resources abound in North East India. The NER is home to approximately 18% of the total hydrocarbon reserves of India. A total of 2,246.6 MMT has been excavated so far. The upper Assam shelf basin has 6,001.2 MMT of hydrocarbon resources and the Assam–Arakan fold belt basin has 3,180 MMT hydrocarbon resources. Tables 2.1, 2.2, and 2.3 show the estimated coal, oil, and natural gas reserve and production in the North Eastern States.[1, 2,3]

2.1.1 North East India's Power Potential and Installed Capacity

In the North Eastern Region, the total installed capacity was 4,896.14 megawatts (MW) as of 28 February 2021,[4] consisting of 2,582.98 MW thermal, 1,944.00 MW hydro and 369.17 MW renewable energy sources (Figure 2.1). Assam has the highest installed capacity for power generation, accounting for 1,794.10 MW, followed by Arunachal Pradesh, accounting for 765 MW of energy (Figure 2.2). Although this power generation is dominated by gas power stations, development of alternate and clean sources, such as hydro

TABLE 2.1 Coal reserved (million tons)

State	Proved	Indicated	Inferred	Total
Meghalaya	89	17	471	576
Assam	465	57	3	525
Nagaland	9	0	402	410
Sikkim	0	58	43	101
Arunachal Pradesh	31	40	19	90

Source: Coal Reserves in North East India, as on 30 April 2021

TABLE 2.2 Natural gas production in North East India (MMSCM)

State/ Region		2013– 2014	2014– 2015	2015– 2016	2016– 2017	2017– 2018	2018–2019 (April to Dec. 18)
Assam	OIL	2,049	2,509	2,618	2,693	2,659	1,895
	ONGC	459	449	405	434	508	366
	Pvt/JV	-	-	-	-	53	230
	Total	2,868	2,958	3,023	3,127	3,219	2,491
Arunachal	OIL	19	12	12	12	12	11
Pradesh	Pvt/JV	22	22	18	16	18	12
	Total	41	34	30	28	30	23
Tripura	ONGC	882	1,140	1,332	1,430	1,440	1,144
North East	Grand Total	3,731	4,131	4,385	4,585	4,689	3,658

Source: Annual Report, MoPNG

TABLE 2.3 Crude oil production in North East India (MMT)

State/ Region		2013– 2014	2014– 2015	2015– 2016	2016– 2017	2017– 2018	2018–2019 (April to Dec. 18)
Assam	OIL	3.445	3.405	3.219	3.250	3.367	2.515
	ONGC	1.263	1.061	0.965	0.950	0.975	0.753
	Pvt/JV	-	-	-	-	0.003	0.024
	Total	4.708	4.466	4.184	3.278	4.345	3.293
Arunachal Pradesh	OIL	0.021	0.007	0.006	0.008	0.007	0.006
	Pvt/JV	0.09	0.069	0.051	0.048	0.043	0.0026
	Total	0.111	0.076	0.057	0.056	0.050	0.032
North East	Grand Total	4.819	4.542	4.241	4.256	4.395	3.324

Source: Annual Report, MoPNG

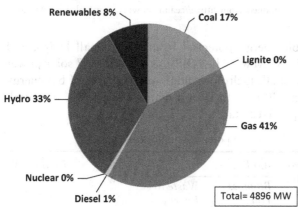

NORTH-EAST REGION POWER GENERATION MIX

Renewables 8%
Coal 17%
Lignite 0%
Hydro 33%
Gas 41%
Nuclear 0%
Diesel 1%
Total= 4896 MW

FIGURE 2.1 Power generation mix of North East India.

Source: Author Graphics

power plants and energy through renewables is also being facilitated, with renewables currently contributing about 8% of the total installed capacity of power generation in India's NER.

2.1.1.1 Renewable Energy

The NER also has huge renewable potential in the form of small hydro power, wind and solar power, but currently these are installed with very minimum capacities. According to the Ministry of New and Renewable Energy (MNRE)'s Annual Report 2020–2021, the North Eastern Region's

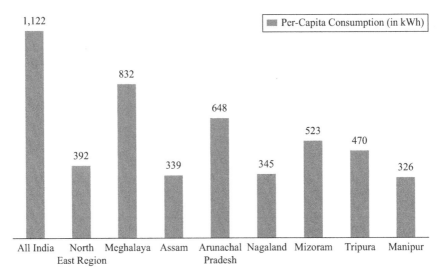

FIGURE 2.2 Installed capacity in North East India.

Source: Author Graphics. Data obtained from https://cea.nic.in/wpcontent/uploads/pdm/2020/12/growth_2020.pdf

total estimated renewable energy potential from solar, small hydro, and bio energy is roughly 65,837 MW.[5] The NER has 62,300 MW solar power potential, 3,261.49 MW small hydro potential and 276 MW of bio energy (Table 2.4). It has around 300–500 MW estimated wind power potential, but this has not been exploited so far.

TABLE 2.4 State-wise renewable energy potential of North East India

States	Small Hydro Power (MW)	Bio Energy		Solar (MW)	Total (MW)
		Biomass Power (MW)	Waste to Energy (MW)		
Arunachal Pradesh	2,064.92	8		8,650	10,723
Assam	201.99	212	8	13,760	14,182
Manipur	99.95	13	2	10,630	10,745
Meghalaya	230.05	11	2	5,860	6,103
Mizoram	168.9	1	2	9,090	9,261
Nagaland	182.18	10		7,290	7,482
Sikkim	266.64	2		4,940	5,209
Tripura	46.86	3	2	2,080	2,132
Total	3,261.49	260	16	62,300	65,837

Source: Annual Report 2021–2022, Ministry of New and Renewable Energy (MNRE), GoI URL: https://mnre.gov.in/img/documents/uploads/file_f-1671012052530.pdf, accessed on 25 May 2021

TABLE 2.5 State-wise installed capacity of grid-connected renewable power

States	Small Hydro Power	Bio Power	Solar Power	Total Capacity	Capacity Addition During 2020–2021
Arunachal Pradesh	131.105	-	11.23	142.34	1.07
Assam	34.11	2	68.57	104.68	17.01
Manipur	5.45	-	12.2	17.65	0.81
Meghalaya	32.53	13.8	4.13	50.46	0.28
Mizoram	36.47	-	7.88	44.35	0.9
Nagaland	30.67	-	3.04	33.71	0.13
Sikkim	52.11	-	4.65	56.76	2.71
Tripura	16.01	-	14.87	30.88	1.31
Total	338.46	15.8	126.57	480.83	24.22

Source: Annual Report 2021–2022, MNRE, Government of India. URL: https://mnre.gov.in/img/documents/uploads/file_f-1671012052530.pdf, accessed on 25 May 2021

However, the grid-connected installed capacity of renewable energy in North East India is only about 421.35 MW. With a capacity of 338.46 MW, small hydro has the maximum installed capacity. After that is solar, with 67.09 MW, and bio energy with 15.80 MW (Table 2.5). Arunachal Pradesh has the highest small hydro installed capacity potential, of 131.10 MW, followed by Sikkim (52.11 MW) and Mizoram (36.47 MW). With 42.99 MW, Assam has the highest solar installed capacity followed by Tripura (9.41 MW) and Arunachal Pradesh (5.16 MW). Bio energy is spread mainly in Meghalaya (13.80 MW) and Assam (2.0 MW).

2.1.2 Power Supply Position in the North Eastern Region

The North Eastern grid faced an energy shortage of 1.4% in February 2021[6] and a peaking shortage of 0.6 during the year 2020–2021 as compared to energy and peaking shortage of 1.7% and 2.4%, respectively, during 2019–2020. However, in February 2021 the peak power demand was 2,680 MW but it met only 2,664 MW and thus there was a shortage of 16 MW, i.e. 0.6%. From April 2020 to February 2021 the peak demand not met was 187 MW, i.e. a shortfall of 5.7% (Table 2.6.).

At the moment, the most serious issues facing the region include, among other things, insufficient generating capacity, an asymmetrical thermal-hydro mix of available generating capacity, transmission barriers, which are primarily caused by the ineptness of the transmission network in the region, the absence of indispensable transmission links, and lengthy outages of crucial elements in the network.

Extensive work has been done in the region to consummate the cherished dream of the Government of India to provide 24 × 7 power to every

TABLE 2.6 Power supply position of North East India

States	February 2021				April 2020 to February 2021			
	Peak Demand	Peak Met	Demand Not Met		Peak Demand	Peak Met	Demand Not Met	
	MW	MW	MW	%	MW	MW	MW	%
Arunachal Pradesh	156	144	12	7.7	158	149	9	5.6
Assam	1,485	1,485	0	0.0	2,072	1,987	85	4.1
Manipur	231	231	0	0.0	252	249	3	1.1
Meghalaya	377	377	0	0.0	384	384	0	0.0
Mizoram	132	132	0	0.0	132	132	0	0.0
Nagaland	152	149	3	2.0	160	155	5	2.9
Tripura (Provisional)	241	241	0	0.0	317	315	2	0.5
North Eastern Region	2,680	2,664	16	0.6	3,294	3,107	187	5.7

Source: Executive Summary on Power Sector February 2021, CEA. URL: https://cea.nic.in/wp-content/uploads/executive/2021/03/executive.pdf, accessed on 25 May 2021

household. As per the Saubhagya website, all seven North Eastern States have completed 100% electrification for every household in the region. Though the generation capacity has increased, there is still a marginal shortfall. Also, when compared to the national average, the per capita consumption in the NER is lower (Figure 2.3).

2.2 North Eastern Regional Grid

The NER's grid is synchronized (Figure 2.4) with the All India Grid via the Eastern Regional Grid through A/C links: 400 kV Bongaigaon–New Siliguri D/C and 400 kV Bongaigaon–Alipurduar D/C and 220 kV Birpara–Salakati D/C. The NER is connected to the Northern Region Grid through +/– 800 kV Multi-terminal Biswanath Chariali–Alipurduar–Agra HVDC link which may carry 6,000 MW. The NER is connected to Bhutan Power System through 132 kV Salakati–Gelyphu (Bhutan) S/C and 132 D/C and to kV Rangia–Motonga (Bhutan) S/C. The NER is connected to Bangladesh through 132 kV Surajmaninagar–South Comilla (Bangladesh) and with Myanmar through 11 kV Moreh–Tamu (Myanmar) S/C.

Having a large number of tripping and overloading in multiple transmission links is a serious concern. Therefore, appropriate strengthening of transmission systems and an improved availability of the telemetry system are required, which is insufficient in the majority of North East Indian states.

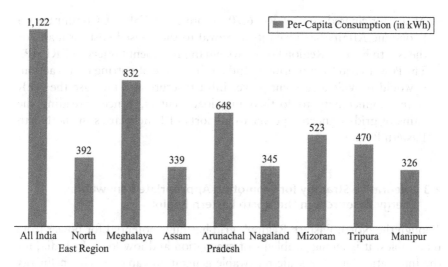

FIGURE 2.3   Per capita electricity consumption (state-wise).

Source: Based on data from Growth of electricity sector in India from 1947–2019, Central Electricity Authority May 2019. URL: https://cea.nic.in/old/reports/others/planning/pdm/growth _2019.pdf, accessed on 13 March 2021

For the improvement of the power sector in the North East and to further facilitate power transfer within the region and cross-border, multiple initiatives have been taken by the NER, which are

- SAMAST (Scheduling, Accounting, Monitoring, and Settlement of Transaction). Metering at an intra-state level, intra-state scheduling, an Automatic Meter Reading (AMR) scheme to get data from State Load Dispatch Centre (SLDC), an open access web portal and other things would be part of its implementation. This would enable the use of intra-State ABT.
- For Inter-State Transmission System (ISTS) Communication, Central Electricity Regulatory Commission (CERC) has published the Communication Regulation that envisions the Centralized Supervision System. The Central Transmission Utility (CTU) is the Nodal Agency for the supervision of the inter-State communication system and will establish centralized supervision in order to quickly discover and restore faults.
- Better weather forecasting for the power sector by the Indian Metrological Department – North Eastern Regional Load Despatch Center (IMD–NERLDC). The IMD would install 150 automatic weather stations (AWS) in the NER and its substations and will share data with NERLDC and SLDCs which will be beneficial for load forecasting.

- At an anticipated cost of Rs. 6,700 crores, the Cabinet Committee on Economic Affairs has given its approval to the revised cost estimate for the North Eastern Region Power System Improvement Project (NERPSIP). The Power Grid Corporation of India is in charge of putting it into action. It would provide a resilient power infrastructure and increase the NER states' connectivity to forthcoming load centers, hence spreading the gains of grid-connected power to all sorts of beneficiaries in the North Eastern Region.[7]

2.3 Integrative Strategy for Promoting Appropriate Renewable Energy Resources in the North Eastern Region

Though the North Eastern Region's grid has huge potential for power from renewables, it is facing problems of evacuation and low load. In India, for the integration of large-scale renewable generation capacity, Green Energy Corridors (GEC), forecasting of Renewable Energy (RE) generation, establishment of Renewable Energy Management Centers (REMC), etc., are being adopted. However, implementation at the state-level in the NER has been unsatisfactory.

TABLE 2.7 Transmission Infrastructure developed for cross-border energy trade

Sr. no.	Name of Transmission Line	Capacity Transmitted (MW)	Voltage Level (kV)	Status of Completion
1	Kurichu (Bhutan)–Gelephu (Bhutan)–Salakati (Assam)	60 MW	220 kV D/C	Completed
2	Dagachhu–Tsirang–Rurichhu–Chukha (Bhutan)	126 MW	220 kV S/C	Completed
3	Surajmaninagar (Tripura) in to Comilla (Bangladesh)	160 MW	400 kV D/C	Completed
4	Moreh (Manipur) to Tomu(Myanmar)	3 MW	11 kV A/C	Completed
5	LILO of Bongaigaon–Siliguri at Alipurduar	1,000 MW	400 kV D/C	Under Construction
6	LILO of BishwanathChariyali–Agra HVDC Bi-pole line at Alipurduar	6,000 MW	800 kV	Under Construction
7	Bornagar (Assam) with LILO of Balipara–Bongaigaon to Bangladesh	2,000 MW	765 Kv D/C	Planned

Source: Author Tabulation

TABLE 2.8 Inter-regional exchange (ER/NR–NER) during 2018–2019 (All figures in MU) (Note: (–) Export to ER/NR I (+) Import from ER/NR)

Month	ER-NER	ER-1\'ER	1\R-NER	Net Export/ Import	Net Deviation
	Schedule	Actual	Actual	Actual	
Apr-18	138.29	240.80	−93.45	147.35	9.07
May-18	−39.57	77.97	−191.36	−113.39	−73.82
Jun-18	16.91	271.66	−309.48	−37.82	−54.72
Jul-18	−44.78	370.48	−481.45	−110.96	−66.19
Aug-18	33.37	518.58	−514.05	4.52	−28.85
Sep-18	10.82	481.10	−492.08	−10.99	−21.81
Oct-18	−76.41	397.73	−496.12	−98.40	−21.99
Nov-18	−11.44	367.99	−387.31	−19.32	−7.89
Dec-18	76.65	−127.72	196.86	69.14	−7.51
Jan-19	72.17	−363.93	446.35	82.42	10.26
Feb-19	157.58	−288.33	414.91	126.58	−31.00
Mar-19	151.74	−322.26	416.25	94.00	−57.74
Total	485.33	1,624.07	−1,490.94	133.13	−352.20

Source: NERLDC, Annual Compendium 2018–2019. URL: www.nerldc.org/wpcontent/uploads/Web_Retrospect_2018-19.pdf, accessed on 15 April 2021

In order to tap the hydro potential in the North Eastern Region an outline of a transmission system has been made for evacuation of power from hydro projects of about 50,000 MW in the NER and 15,000 MW in Sikkim/Bhutan. For the T&D project in Arunachal Pradesh and Sikkim, the entire cost is borne by the Government of India (GOI) through the plan scheme of Ministry of Power. The remaining six states, namely Meghalaya, Mizoram, Assam, Tripura, Manipur, and Nagaland, would be funded equally by the World Bank and the Government of India through the budget of the Ministry of Power. The "NER power system upgrading project," financially supported by the World Bank, requires the power grid to carry out transmission operations in six states of the NER in tough terrain.

Investments in the evacuation network and grid management systems are necessary for renewable energy grid integration. Many countries' experiences show that grid networks can absorb a high amount of RE through improvements in power system operations and regulatory frameworks and market reforms. The Central Electricity Authority (CEA) of India predicts that renewable energy would account for 17.5% of the country's electricity grid till the end of 2022, depending on the country's 160 GW of solar and wind power capacity. At these penetration levels of 15% to 20%, grid integration is much easier.

Reliability and cost-effectiveness of power systems in a high-renewable scenario can be ensured by implementing incremental technological and economic grid management solutions based on existing international practices. Nevertheless, wind and solar power variability is the primary technical difficulty in renewable energy integration, as it impacts the load generation balance, the fluctuating requirement for reactive power, and the effect on voltage stability. Conventional power sources are still needed to provide backup power when RE power is unavailable. Renewable energy source (RES) induction doesn't lessen this need. As renewable energy sources become more variable, grid stability and security are put under greater strain.

2.3.1 Renewable Energy Integration in the NER Grid

Hydro and solar power generation are viable options for the NER. Similarly, the NER has tremendous potential for generating small hydro power, but the Small Hydro Policy for the NER was notified in 2007 and has not been modified since. The NER also has significant biomass potential, which can be utilized for the establishment of a power plant at the grid's tail end. However, a lack of familiarity with the concept, financing, and PPP structure is preventing other people from pursuing similar applications.

There have been regulatory frameworks like scheduling and forecasting of RE, and flexible operations of thermal power plants and auxiliary services are put in place by the Central Electricity Regulatory Commission (CERC) that are designed to encourage the growth of RE penetration in India. Increased grid balancing and transmission network improvements are needed in order to attain a bigger share of renewable energy in the NER's system.

Current forecasting methods like Renewable Energy Management Centers (REMC) for real-time monitoring are required in the NER states to improve forecast quality to maintain load-generation balance. To encourage conventional generation sources to be more flexible, regulatory assistance must be increased as well. In research titled "Greening the Grid," Power System Operation Corporation (POSOCO) and the National Renewable Energy Lab (NREL) concluded that integrating 100 GW of solar and 60 GW of wind may be achieved at only 1.4% curtailment without fast-ramping infrastructure like batteries, pump hydro, or gas-based plants. Coal-fired power plants' "inherent flexibility" will aid in the integration of fluctuating renewable energy sources, according to the model. For this reason, it is expected and required that existing plans for increasing generation and transmission capacity, which would allow for the handling of errors in RE projections, changes in net load (ramps), and low RE generation, are carried out in a timely manner.

2.3.2 Gap Analysis for Solar PV and Wind Power Policy

In 2019, the Ministry of New and Renewable Energy, Government of India sanctioned/approved RTS projects of 55.05 MWp capacities to various

North Eastern States, of which 32 MWp had been sanctioned in the 2018–2019 financial year. The NER's solar capacity has expanded significantly. Furthermore, state regulatory commissions have issued net-metering regulations. Manipur and Assam have also announced policies on rooftop solar. No large-scale wind power projects have yet been established in the NER, according to the Indian Wind Atlas, which estimates the region's wind potential at 50 meters to be 406 MW. Investors' willingness to put money into the development of renewable energy sources is being hampered by an absence of information on feed-in-tarrifs (FIT) and capital cost. That the must-run status is being given to only 5–10 MW RE projects is a negative signal for investors and impacts the growth of power from solar and wind energy.

Precision in forecasting and scheduling are necessary for the successful integration of wind energy into the power grid. Solar and wind power variability affects the load generation balance, as well as the shifting need for reactive power, which has a direct impact on voltage and grid stability. Measures to reduce the consequences of fluctuating reactive production of power must be put in place at the pooling station, and plans must be made to support intra-state power evacuation as well. The following are the three distinct issues that have an impact on the grid integration of solar and wind energy:[8]

1. **Variability of resources**: Due to varying wind speeds and solar intensity, power plant operators have no control over the output of wind and solar. Additional energy input and peripheral ancillary services, such as voltage and frequency regulation, are required to instantly balance supply and demand.
2. **Unpredictability**: To a certain extent, wind and solar energy's availability is unpredictable. Electricity is only produced when the wind blows, and the presence of sunlight is required for PV systems to function. Developments in advanced forecasting technology allow for better control of unpredictable events. Standby reserves and dispatchable loads are part of technological systems, which can be used in the event that renewable sources produce less power than predicted or produce more power than predicted, respectively.
3. **Location dependence**: There are only a few places in the world where high-quality wind and solar resources are available for the production of renewable energy, and the places where that energy is ultimately used are often far from those few places. New transmission capacity is needed to connect wind and solar power to the grid. Transmission costs are also higher in hilly areas because of the logistics involved.

2.3.3 Gap Analysis for Hydro Power Policy

The development of hydro power projects in the North Eastern States has enormous potential. It has been a priority for the MNRE to develop small hydro power projects in the North East. The NER's surplus seasonal renewable power must be utilized. Geo-politics is a major concern for most of the river systems in the North East. Hydro power generation is hindered as a result of this. A mechanism must be established by the states and the central government to resolve outstanding border and share allocation issues.

Because of their inherent advantages, such as the ability to start and stop quickly, hydro power plants and storage-type hydro plants are ideal for providing a balancing service. To encourage private investment in the region, small-hydro power (SHP) policy needs to be updated with new incentives and promotional measures.[9]

Studies like potential assessment, as well as risk and impact analysis have been performed for the hydro project in the Himalayan Range. It has been found that exploring and developing the identified projects provides huge benefits to the region and local economy. Pumped hydro storage also holds the key for RE integration in the NER as they provide an effective tool for grid balancing. While there are challenges associated with development of hydro power, considering the benefits of hydro such as load balancing, fuel at zero cost, easy operation, etc., there is definitely a need to pursue its development.

2.3.3.1 Pump Storage

Pumped storage plants assist in providing peak power and preserving the power system's stability. The base load is usually provided by thermal or nuclear power facilities, whereas peaking loads are commonly met by conventional hydro and pumped storage plants. Pumped storage plants can improve thermal station capacity utilization and eliminate operational issues during low-demand periods. The development of pumped storage also provides substantial reactive capacity for regulation and low-cost spinning reserve to handle rapid network load changes.

That the energy gained through pumped storage development is always less than the energy input should not conceal the fact that this system loss is negligible in comparison to the significant fuel savings realized when these stations are managed in an integrated manner.[10] The prospective states include Manipur, Mizoram, and Assam.[11]

2.3.4 Challenges to Grid Integration of Renewable Energy

It is difficult to integrate renewable energy into the grid in India's North East Region, as well as to build an evacuation system. The main problem is to ensure that transmission lines are in place before renewable energy projects are ready, because transmission projects can take up to 5 years to complete, as opposed to solar installations' 12–18 month timeline.

The following is a summary of the findings of this gap across different technologies and parameters based on the gaps mentioned above.

1. Inadequacies in grid infrastructure

   As a result of the lack of an adequate plan to create a dedicated infrastructure for RE evacuation, there are fewer demand centers in the North Eastern States and a lower possibility of earning premiums under open access. The inter/intra state grid is needed to transport the extra power generated in the states to the rest of the country.

   Issues about the North East's grid infrastructure must be addressed. There have been delays and the infrastructural building does not keep pace with tenders coming out despite the fact that the GEC program aims to evacuate power from renewable energy-rich areas to other states via 765 kV and 400 kV high-voltage transmission lines. Tenders are issued in some states without contacting the corresponding "state energy regulatory commission" (SERC); as a result, when PPAs are brought to regulatory commissions for confirmation, they are stalled because the SERC alleges a lack of transmission infrastructure.

2. Forecasting of wind and solar generation

   Because of the critical nature of ensuring the safe operation of the grid, it is critical for the system operator to anticipate what is likely to occur in the next few hours so that proper steps can be taken. For the goal of wind forecasting, a number of companies have joined the field of forecasting, using historical data, air pressure, wind speed, humidity, topological aspects of the region, and other factors.

3. Commercial mechanism implementation

   An appropriate market design to manage reserves for power balancing; flexible generators; and an ancillary market are all important considerations. Performance metrics for control area, formulation of intra-control area deviation settlement mechanism and development of visualization and situational awareness.

2.3.5 Comprehensive Integrative Strategy for Promoting Renewable Energy

The NER has the potential for developing hydro and solar power, but a lack of policy on the solar sector and challenges in the development of the hydro sector are restricting investment in the region.

Solar and wind power are the most underutilized sources of renewable energy in this region, although the region has huge potential for solar power. A broad-scale promotion of hybrid energy systems, which are often composed of two or more renewable energy sources coupled in such a way as to create an efficient system with uninterrupted power supply, can be achieved in the region that has a lot of hydroelectric power potential. But advanced forecast and scheduling need to be done for adequate balancing as these are intermittent and variable sources of energy. As per the Central Electricity Authority (CEA), the balancing power requirement for the Eastern and North Eastern Region is 3,000 MW/hour, which is the lowest among all the regions. Regulatory support must be enhanced to incentivize the flexibility of conventional generation sources, as the existing tariff arrangements and balance sheets of state-owned distribution companies are fundamentally untenable.

2.3.6 Renewable Energy Integration and Grid Stability

The grid's integrity, stability, and security are the primary considerations for integrating renewable energy sources to it. Decarbonizing the power sector while still meeting rising energy demands necessitates the integration of Variable Renewable Energy (VRE) sources like wind and solar PV. However, it has technical and operational difficulties when it is integrated into the grid. To make sure the integration of renewable power, the entire power system and policies are to be designed keeping in mind the following objectives:

- Reliability of generating unit.
- Grid security and stability to supply affordable and reliable power to consumer.
- Minimize the cost of flexible operation.
- Maximize the power generated through renewable energy sources and its integration into the grid.
- Minimize investments through optimal utilization of existing infrastructure and assets.

There is a marked difference between conventional generating plants and RES. Load requirements in conventional plants can be programmed to vary, which is why it is referred to as dispatchable generation. By contrast, renewable energy sources such as solar and wind generation are dependent on nature and thus these types of energy sources are referred to as non-dispatchable. Variability and uncertainty are inherent in RE generation.

Variability and uncertainty in the generation of power from VRE necessitate a reserve of traditional balancing power and energy storage devices to ensure that demand can be satisfied at any moment. Another factor to keep in mind is that solar plants produce the most power during the day, when demand is low, and the least power at night, when demand is high. Conventional generators must be able to ramp up quickly in order to meet this demand. Because of this, it is difficult to integrate VRE sources into the existing grid.

The following factors of RE generation are vital to be considered for RE Integration and grid stability:

- Due to the inherent fluctuation of wind and solar resources, determining whether a system with significant variable renewable energy has enough supply to meet long-term energy demand becomes more complicated.
- Wind and solar power are two renewable energy sources and are non-dispatchable since they rely on nature.
- During the early morning hours, solar generation increases gradually and reaches its peak at about midday. After that, solar generation begins to decline and eventually disappears completely as nighttime approaches. At noontime, solar plants produce the most power, whereas at night they produce nothing.
- It is extremely difficult to anticipate the output of a solar plant due to the fact that cloud movement is highly unpredictable.
- Wind energy is affected by weather patterns on a daily and seasonal basis. When a storm is approaching, changes in wind generation take place gradually over the course of several hours. The wind follows a seasonal pattern, reaching its peak during the monsoon season.

Considering the inherent characteristics of RE generation, the following actions are to be taken for RE integration and grid stability.

1. In order to get better access to the least expensive power generation, neighboring states should coordinate their schedules and dispatches with each other.
2. Incorporation of state-of-the-art automated load and renewable energy forecasting systems.
3. Scheduling and dispatch could be upgraded to 5 minutes from the current 15 minutes basis.
4. CERC guidelines for coal flexibility, reducing minimum operating levels for coal plants.
5. New tariff structure that specifies performance criterion (ramping), and addresses the value of coal as PLF decline.
6. Creation of model PPAs for RE that move away from must-run status and employ alternative approaches to limit financial risks.
7. Address distribution grid integration concerns, such as rooftop PV and utility-scale wind and solar coupled to low-voltage lines.
8. RE generators to provide grid services such as automatic generation control and operational data.
9. Increase access to existing coal, gas turbine, hydro, and pumped storage resources through regulatory and policy measures.
10. CERC/SERC to issue regulations to enable policy-related interventions.

11. CTU/STU should upgrade the technologies and make necessary investments to handle the intermittency through appropriate technical interventions.
12. Availability of substantial flexible generation that can ramp up very quickly.
13. Availability of storage devices as reserve.
14. Procurement mechanism of flexible generation in assuring grid stability.
15. Fast trading of power at power exchange to manage variable generation.
16. Fair price discovery and compensation of flexible resource providers.
17. Expand balancing areas to reduce variability by offering more balancing resources/demand.
18. Evacuation of power from locations with a high concentration of RES via Green Energy Corridors.
19. Upgrade grid operational protocols to prevent renewable energy grid disruption.

2.3.6.1 Balancing Energy Reserves

The demand and supply of power must be equal at all times in a synchronous electrical system. The variation in frequency is a reflection of any imbalance. This means that if there is an imbalance in the system's energy supply, balancing power must be provided.

Balancing power can be classified as Primary Control (PC), Secondary Control (SC), or Tertiary Control (TC) based on their function, response time, and activation method.

Within 30 seconds, the PC could be fully operational. It is activated by a frequency deviation that is measured locally. Hydro power is the primary source of PC. SC could be operational within five minutes of activation. System operators activate it automatically and centrally. SC is used in conjunction with PC to restore frequency and rebalance the respective balancing area. SC can be supplied primarily by hydro power and gas-fired power plants, as well as, to a lesser extent, by synchronized thermal generators. Over time, TC gradually replaces SC. A 15-minute timer can be activated either directly or via a schedule. Standby thermal generators provide the majority of the TC.

With the introduction of VRE into the system, the concept of balancing power has taken on new significance.

Market operations are mostly carried out by power exchanges in India. They must work together in order to implement and balance the system properly. Listed below are the roles of each of the several entities in the integrated approach:

2.3.6.2 Balance Responsible Parties (BRPs)

A BRPs entity, also called "program responsible parties," has the responsibility of balancing a portfolio of generators and/or loads. BRPs can be utility companies, industrial consumers, and so on. They provide system operators with legally enforceable schedules for each quarter-hour of the following day, and they are held financially liable for any departures from these plans.

If the aggregate of BRP imbalances is non-zero, system operators use balancing power to physically balance demand and supply.

2.3.6.3 Suppliers of Balancing Power

Suppliers of balancing power hold supply capacity in reserve and supply energy when the system operator activates it. They are obligated to deliver energy in accordance with pre-determined terms, such as a specific time frame and a specific rate of ramp. Suppliers are typically thought of as primarily generators, but they can also be buyers.

When balancing power is deployed, a system is said to be "actively balanced." The capacity and energy costs for balancing electricity are included in the pricing of this service.

Similarly, a system is said to be "passively balanced" if the Balance Responsible Parties (BRPs) take the price signal and balance it.

There are two ways in which balancing power can be ordered: either through contracted power or through sending imbalance price signals to BRPs and passively balance the system. "Self-balancing" is another name for this phenomenon. It can take several minutes for passive balancing to take effect. Therefore, it cannot be used as a substitute for the balancing power needed to respond to stochastic shocks.

In a high VRE infusion scenario, finding the balance reserve can be a challenge. There is no one-size-fits-all approach to this. Rather than using a deterministic technique, many professionals now opt for a stochastic approach. Forecast mistakes for wind and solar generation are discovered using historical data. It is assumed that the errors have a normal distribution. The reserve requirements are estimated using a confidence level of 95 or 99%.

2.3.6.4 Advance Forecasting

Variability in solar generation is primarily caused by clouds. When forecasting the next few hours, satellite photos can be utilized to determine the path of approaching clouds. Weather models can be used to predict the formation and evolution of clouds over longer timescales.

Using forecasts for wind and solar power can help reduce the risk of renewable energy generation.

By being able to predict wind and solar power variations more accurately, grid operators can better plan for extreme events, such as those that have exceptionally high or low levels of renewable generation. However, there is a requirement for SAMAST implementation at the intra-state level.

2.3.6.5 Effectiveness of Market

Flexible resources are affected by the design of the market. Pricing, schedule/dispatch interval, ancillary service demand, capacity market, and other approaches can all help meet the variable renewables' flexibility needs.

A regional/national balancing market mechanism is needed, and power generators are concerned about ramping up or down, minimum turn down, and hot start up.

2.3.6.6 Demand Response

Customers are encouraged by demand-side management methods to use as much variable renewable energy as possible while the supply is abundant. When wind and solar PV are generating more power than demand, demand response can encourage consumers to use more electricity through appropriate price signals of low rates, thus shifting the load and ensuring that generation resources are better utilized.

2.3.6.7 Flexibility

Flexibility in the power system has evolved over time in response to changes in technology and the power market. It refers to the ability to function reliably with considerable amounts of variable renewable energy.

According to the International Energy Agency (IEA) (2011): "Flexibility expresses the extent to which a power system can modify electricity production or consumption in response to variability, expected or otherwise."

The IEA (2014) introduced a distinction between a broader concept of flexibility and a narrower concept of ramping flexibility: "In a narrower sense, the flexibility of a power system refers to the extent to which generation or demand can be increased or reduced over a timescale ranging from a few minutes to several hours."

2.3.6.8 Minimum Operational Level of Generation

The flexibility from a thermal generator is technically constrained by its minimum operational level. Reducing the lower limit of operation will allow plants to stay in service during periods of low demand or high RE generation, thereby reducing the number of start–stop cycles. Recognizing this, the Indian Electricity Grid Code has been amended by the CERC in the 4th amendment by mandating the technical limit of thermal stations to 55% of installed capacity. Clause 6.3B (1) of IEGC is reproduced below.

"The technical minimum for operation in respect of a unit or units of a Central Generating Station of inter-state Generating Station shall be 55% of MCR loading or installed capacity of the unit of at generating station."

Further the CEA Standard Technical Specification for Sub-critical & Supercritical Thermal Power Project requires the thermal plant to be stable at 40% of load without any oil support. Relevant clauses have been reproduced below.

The design of steam generator shall be such that it does not call for any oil support for flame stabilization beyond 40% Boiler maximum continuous rating (BMCR) load when firing any coal from the range specified, with adjacent mills in service and mill load not less than 50% of its capacity.

2.3.6.9 Minimum Load

The lowest possible net load a generating unit can deliver under stable operating conditions is referred to as Minimum Load. It is measured as a percentage of normal load or the rated capacity of the unit.

2.3.6.10 Start-Up Time

The period from starting plant operation until it reaches its minimum load is called the start-up time. It varies greatly with different generation technologies. Other influencing factors include down time (period when the power plant is out of operation) & the cooling rate.

2.3.6.11 Start-Up and Shut Downs

The start-up time is defined as the period from starting plant operation until minimum load is attained. In the Indian context, time taken for start-up is declared by its owner. Shut-down time is linked to the ramping down capability of the stations.

2.3.6.12 Minimum Thermal Load (MTL)

The MTL is the ratio of actual minimum load on the prime mover of a thermal power station and its rated capacity. E.g. if a 200 MW plant runs at minimum load of 120 MW during a day, then the MTL for that plant is 120/200, i.e. 60%.

2.3.6.13 Flexibility Associated with Conventional Generating Units

To accommodate the variability and uncertainty of generation from RES, the conventional generating plants must be flexible. In terms of flexibility, hydro plants, pumped storage plants, open cycle gas turbine, gas engines, etc. are very suitable.

2.3.6.14 Flexibility Indices

Electricity demand is continuously varying due to weather and several other factors. Electricity generation shall adjust to load requirements for maintaining load generation balance at all times. Since RE is variable in nature, integrating them into grid further increases variability requirement from other generation sources.

2.3.6.15 Ramp Rate

When a power plant is in operation, its ramp rate measures how quickly it may alter its net power output. It can be mathematically expressed as a change in net power, ΔP, per change in time, Δt. Ramp rate is usually specified in MW/min or percent of the rated load per minute (percent P/min). Generally, ramp rates are influenced by the technology used to generate the power. Ramp rate is one of the important indicators in determining the flexibility behavior. For integration of high RE, system-wide ramping would be a key parameter.

2.3.6.17 Transmission Strengthening

Transmission networks are required to deliver the renewable power produced by the states to the load centers on the grid. This may necessitate the expansion of existing grid networks, or it may necessitate the usage of storage devices instead. Green Energy Corridors for the evacuation of power from locations with a high concentration of renewable energy sources are now being developed in India.

2.3.6.18 Regulatory Initiatives

A variety of regulatory efforts have been taken by CERC in support of RE development and further to facilitate effective integration of RE in the grid. The following is noted from the order dated 12 July 2020 against Petition No. 269/MP/2019.

> The Commission has taken a number of regulatory initiatives for the development of RE sources and for the smooth integration of RE in the grid. In order to facilitate effective integration of variable and uncertain RE generation, the Commission has specified Roadmap for Reserves,

framework for Ancillary Services Operation besides amending the IEGC, which provides for technical minimum of 55% in case of thermal generating units aimed at providing flexibility to respond to the needs of variation in demand, RE generation etc. Besides the above, the Commission has also brought in several regulatory interventions for promoting renewable energy generation, which inter alia include, notification of Renewable Energy Certificate Mechanism, Framework for Scheduling, Forecasting & Deviation Settlement of RE generation, specifying Relaxation in Deviation Settlement Mechanism for RE generation, etc.

2.3.6.19 *Flexibility of Thermal Generation for Renewable Integration*

POSOCO's "Flexibility Analysis of Thermal Generation for Renewable Energy Integration in India" found that the vast majority of coal-fired power plants in the country are flexing their generation to the tune of 20–30% of their installed capacity; 60% of units across the country have granted flexibility in the range of 20–30% of IC.

I. At a national level there are only 20% of units flexing their generation above 30% of IC.
II. The introduction of a technical minimum of 55% at the central level is a significant component in gaining greater flexibility. If the minimum operational level of state-controlled thermal generating stations are also reduced, greater flexibility would be achieved aiding higher RE integration.
III. Region-wise and all-India average, maximum, minimum, and flexibility of thermal generation are all given in Annexure – A, indicating growth and seasonality behavior.
IV. The CERC has taken major initiatives, such as reducing the technical minimum to 55% and mandating 1% ramp rates, etc., in the direction of improving flexibility in the Indian power system. To integrate ever-increasing levels of RE, the regulatory framework at inter-state and intra-state level has to be amended in line with global standards in terms of minimum operational levels of generation, two shifting operations, etc. Suitable incentives also need to be framed for provision of flexibility. Technological upgradation of units needs to be encouraged for achieving higher level of flexibility.
V. The reduction in technical minimum turn down level from the present (55% DC on bar) to a lower value can be a good technological intervention to take up. A greater plant load factor may still be achieved by low-variable-cost thermal power plants in India, despite the country's increasing solar penetration rate. The mid merit order power plants are required to do more ramp up and ramp down duty.

The flexibility of a power plant is measured by the ratio of the difference between its maximum and minimum generation in a day to its total installed capacity. For each unit, a calculation of its flexibility was made on a daily basis. The average flexibility of the units is calculated by averaging the day-to-day flexibility of the units over the entire operating period.

To summarize, given India's enormous renewable energy potential, the country has set a capacity augmentation target of 175 GW by 2022. The following considerations must be kept in mind in order to achieve the integration and RE generation into the grid:

1) Equipment for renewable energy development must be manufactured in a way that allows for rapid development. The Solar Energy Corporation of India (SECI) has previously granted 4 GW of RE projects linked to the establishment of manufacturing facilities in India to encourage the establishment of RES-related equipment manufacturing facilities in India, in line with the Government of India's "Make in India" strategy.
2) To evacuate RE generation associated transmission system to be planned and its availability to be synchronized with the target commissioning of RE generating stations and to ensure grid integration and stability.
3) The repowering of existing projects on wind power should be encouraged in tandem with the construction of new wind power projects in order to increase the CUF and generate more power from the same locations.
4) The adoption of new technology, as well as mechanisms for the availability of round the clock (RTC) and minimum specific period of hybrid RE generation should be encouraged.
5) Hydro and gas will play a significant role in India, but coal is likely to offer most of the flexibility. There are many ways to relieve the constraints of large-scale renewable generating integration, including flexible operation of coal, hydro, and gas plants, and demand-side measures.
6) Given the current high prices of storage technologies, approximately 1% annual curtailment of renewable energy could be allowed for grid security and stability; in the long run, RE curtailment should be discouraged.
7) A financial framework must be established in order for thermal power plant operators across the country to adopt flexible operation.

2.4 Cross-Border Energy Trade between the NER and South East Asia (ASEAN)

The South East Asia region has abundant hydro power and solar power sources, which, if utilized and shared, can help meet the growing energy demand within the region. In order to tap into these resources and establish transmission lines for the transfer of power between countries, a great deal of work must be done. Cross-Border Electricity Trade (CBET) will be a valuable source of foreign exchange for nations with an abundance of power, while providing chances for countries with a shortage of electricity to rejuvenate their manufacturing sectors by making electricity more affordable, more readily available, and more secure.

Though there is substantial cross-border power trade between India and South Asia (Nepal, Bhutan, and Bangladesh) is going on, South East Asia, especially Myanmar, provides an ample opportunity for cross-border power trade.

2.4.1 India–Myanmar Power Trading

Nagaland, Mizoram, Manipur, and Arunachal Pradesh are all Indian states that border Myanmar. From the north to the south of the Himalayas, there are numerous mountain ranges that supply the perennial rivers. The Irrawaddy, Salween (Thanlwin), and Sittaung rivers are separated by the mountain chains of Myanmar's three river systems.

According to the Ministry of Electricity and Energy, by April 2019, 43% of the total households have been electrified which constitutes 4.67 million households. Still 6.2 million households don't have access to reliable electricity. The National Electrification Plan of Myanmar aims to achieve 100% electrification by 2030.

One-eleventh of the world's average annual usage of electricity is consumed by the average person each year. Since the Myanmar Investment Law came into effect, the power sector has received roughly $21.2 billion in foreign direct investment (FDI), or 27% of the total FDI allowed in the country. As a result, improvements in the power industry, such as increasing electrification rates, could help the economy.

2.4.1.1 Energy Scenario

Myanmar is rich in both onshore, especially the coastal region, and offshore hydrocarbon reserves. Myanmar exports about 90% of its natural gas production to countries such as Thailand, India, and China. The Oil and Natural Gas Corporation (ONGC) is actively exploring the gas in the region and connecting it though pipelines all the way to Tripura. The pipeline connects Tripura with the rest of India and enables it to supply both imported gas from Myanmar and domestic gas to the thermal and gas-based power plants in India. Myanmar's proven oil reserves are estimated at approximately 50 billion barrels (US Department of Commerce, 2016). There are 283 billion cubic meters of confirmed natural gas in the gas reserves. About 500 million tons of coal are stored in Myanmar's coal mines.

2.4.1.2 Power Scenario in Myanmar

Myanmar has a current installed generation capacity of 6,100 MW. Hydro power constitutes around 60% of the installed capacity, while natural gas and coal accounts for 35% and the remaining 5% is accounted for by renewables including small hydro and solar. A total of 3,750 MW of hydroelectric power is now being generated by 27 state-owned hydroelectric projects. A staggering 15 GW of additional electricity is predicted by the Japan International Cooperation Agency (JICA) for Myanmar by 2030. The Ministry of Electricity and Energy expects to generate 8,896 MW, 7,940 MW, 4,758 MW, and 2,000 MW from hydro, renewable energy, coal, and gas power plants by 2030–2031.

Electricity usage in 2015 was 2,527 MW, according to the National Electricity Master Plan (NEMP), which predicts an increase in demand of 4,530 MW by 2025, 8,121 by 2025, and 14,542 by 2030. This increase in demand is due to the government's commitment to 100% electrification by 2030. Transmission networks of 66 kV, 132 kV, and 220 kV power the entire country via the national grid. At present, there is a single 11 kV transmission line between Moreh (Manipur) to Tomu (Myanmar) to trade 5–6 MW power from Manipur to Myanmar.

2.4.1.3 Renewable Energy in Myanmar

The estimated overall potential of Myanmar's hydro resources is approximately 100 GW. The government of Myanmar encourages the use of renewable energy and private sector involvement in the electrical industry.

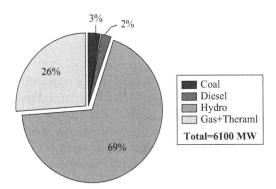

FIGURE 2.4   Power mix in Myanmar.

Source: Based on data from the Ministry of Electricity and Energy (MOEE), Myanmar. Ministry of Electricity and Energy Myanmar (2019) URL: https://greatermekong.org/sites/default/files/Attachment%2011.3_Myanmar.pdf, accessed on 14 April 2021

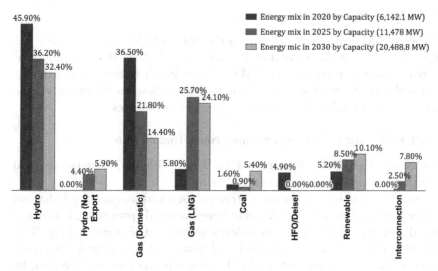

FIGURE 2.5 Projected energy mix of Myanmar.

Source: Based on data from the Ministry of Electricity and Energy (MOEE), Myanmar. Ministry of Electricity and Energy Myanmar (2019) URL: https://greatermekong.org/sites/default/files/Attachment%2011.3_Myanmar.pdf, accessed on 14 April 2021

Renewable energy sources like solar, wind, biomass, and geothermal power generation are slated to account for 2,000 megawatts of the country's total electricity supply capacity by 2030, according to the JICA's preliminary Power Generation Development Plan.

Mandalay, in particular, has a large amount of solar power in Myanmar. There are presently six projects in the works. A total of 470 MW of power has been purchased from three projects, all of which have completed financial close. The International Finance Corporation, the UK Department for International Development, and AusAID are also providing financial support for three floating solar power projects, each with a capacity of 30 MW, that are being developed by independent power producers.

NEMP reports that there are 1,460 MW of solar PV projects in the pipeline, including 150 MW at Nabuaing, 150 MW at Wundwin (Meikgtila), 880 MW at Sagaing, Mandalay, 170 Wundwin (Min Bu), 100 Wundwin (Thapyaysan), and 10 Wundwin (Shwemyo). NEMP also wants to put up a total of 90 MW of floating solar power. Among the upcoming wind power projects are 3,648 MW in Chin, Rakhine, and Yangon, 830 MW in Rakhine, Ayeyarwaddy, and Yangon, 1,000 MW in Shan, Kayah, 1,000 MW in Tanintharyi, Mon, and Kayin, and 30 MW in Chaung Thar.

2.4.2 India–Myanmar Interconnection

Presently 3 MW of power supply runs through an 11 KV line from the 33/11 KV sub-station at the international border town Moreh, India, with a dedicated transformer capacity of 5 MVA. Numaligarh Refinery is collaborating with various parties in Myanmar to deliver diesel to Myanmar via the Moreh–Tamu border, thereby lowering logistical expenses.

2.4.3 North East India as an Emerging Power Trading Hub

As North East India is bordered by Nepal, Bhutan, Bangladesh, and Myanmar it can act like a power trading hub.

Assam lies in one of the most strategic locations, connecting India and other North Eastern States. Despite huge potential from solar, hydro and wind power generation, Assam is deficit in power and energy supply. With the integration of a regional grid and enhancing cross-border interconnection, Assam has the potential to become a power hub in the region, by utilizing its already developing transmission infrastructure and generating revenue from sale/purchase of power.

Arunachal Pradesh has largest potential from hydro power generation among all North Eastern States, a lot of which is unutilized. Arunachal shares border with Bhutan, Myanmar and Assam. Currently, Arunachal Pradesh is a power deficit state, once the hydro projects under construction gets commissioned, Arunachal will become a power surplus states, supplying power the other North Eastern States. For reliable power supply from Arunachal to other states as well as countries, massive transmission infrastructure and investment is required.

Manipur is a power deficit state, but with the upcoming projects Manipur will become a power surplus state in coming years. Manipur has huge untapped hydro potential, and shares border with Myanmar and Assam. Manipur has huge scope for cross-border trade of Power, but issues like isolation, difficult geographical terrain are major challenge for power sector development.

Meghalaya is a power deficit state, with a deficit of 2 MW and 1 MU in 2018–2019. The state has huge hydro potential, and shares border with power hungry region like Bangladesh and Assam. The present transmission links for cross-border energy trade passes through Meghalaya. Due to its strategic location, the state of Meghalaya can be developed as a transit state to transfer power from different border regions.

Mizoram is also a power deficit state. It shares a border with two of the most power-hungry and energy-poor countries, Bangladesh and Myanmar. Mizoram has huge potential for power generation from hydro and solar, most of which is untapped.

One of the most power-starved states in North East India is Nagaland. The state's ability to generate energy is enormous for hydro and solar, of which only 3% has been tapped. Nagaland shares a border with Assam and Arunachal Pradesh in India, and with Myanmar. Massive infrastructure for power transmission and generation is required in Nagaland to fill the supply–demand gap in the state. CBET holds immense opportunity for power sector development in the state.

Sikkim is a power surplus state. The bulk power generated in the state is transmitted to other North Eastern states through regional grid and inter-state transmission system. Sikkim shares a border with Nepal and Bhutan. The state has huge hydro and solar potential, out of which only 10% has been developed. CBET holds immense opportunity for Sikkim to act as a power hub in the region.

Tripura is a power deficit state. The state shares a border with Bangladesh on three sides. It currently transfers power to Bangladesh and Nepal. The state has hydro potential of 15 MW, none of which has been tapped, and also holds huge potential from solar power generation.

2.5 Conclusion

North East India is an emerging powerhouse for ensuring India's sustainable and green energy security. Nearly 40% of the country's entire hydro power potential can be found in this region. As of 1 July 2020, 1,727 MW (or 2.92%) has been harnessed. It is planned and in action to build an additional 2,300 MW of hydro power.[12] It has also huge potential for natural gas and coal. Around one-fifth of the country's hydrocarbon reserves are in the North East, which has yet to be fully explored. It could be used by the government to fulfill the requirements of bordering countries demand. There is a lot of potential for solar power in the North East of India, but the total installed capacity is barely 17 MW, while 85 MW is in pipeline. New viability gap funding (VGF) has been given by the MNRE to help with the development of solar power in the North East States. The government wants to support these projects with VGF funding of up to INR 10 million per MW. The North East States have around 300–500 MW wind power potential, but this has not been exploited so far.

If all the energy potential is harnessed, the North Eastern Region can become an energy hub and may facilitate cross-border power trading.

Large-scale renewable power generation is only possible if there is a comprehensive renewable energy grid integration in North East India. Increased grid balancing and transmission network improvements are needed in order to attain a bigger share of renewable energy in the NER's system.

In order to maintain a load-generation balance, the NER states must have access to cutting-edge forecasting systems like Renewable Energy Management Centers for real-time surveillance. Regulatory support is also needed to be improved to incentivize flexibility of conventional generation sources because the tariff framework in place and the balance sheets of state-owned distribution companies are fundamentally unviable.

It is important that renewable energy integration, the entire power system, and policies are designed while maintaining the reliability of generating units and grid security. A stable supply of affordable and reliable power to the consumer can be achieved through optimal usage of available assets and infrastructure, while maximizing renewable energy generation and integration into the grid.

While deficit countries have possibilities to reinvigorate their manufacturing industries as a result of better affordability and availability, cross-border power trade will become a significant source of foreign exchange revenue for the surplus countries.

Coordination across Bangladesh, Bhutan, Nepal, Myanmar, and India's North Eastern Region can reduce production costs. As a result, it will enable the generation of export revenues and the growth of the regional economy through consumption spending.

Though presently 3 MW of power supply flows between the North Eastern Region and Myanmar through a 11 KV line from the 33/11 KV sub-station at the international border town of Moreh, India, with a dedicated transformer capacity of 5 MVA, it is expected to increase in the near future, and will ultimately go to other South East Asian countries. The Government of India is also working on the "One Sun, One World" project, which promotes the grid connectivity of North East India to many South Asian countries.

Additionally, Numaligarh Refinery is collaborating with various parties in Myanmar to deliver diesel to Myanmar via the Moreh–Tamu border, thereby lowering logistical expenses associated with carting petroleum products from Yangon to Myanmar's north eastern regions.

A harmonious regulatory regime, dedicated network of transmission lines, uniform grid code, harmonious tariff fixation, continuous investment flow, rigorous energy diplomacy, and strong political will can enhance the renewable energy integration and cross-border power trade further.

Notes

1 www.coal.nic.in/content/coal-reserves.
2 http://petroleum.nic.in/sites/default/files/AR_2018-19.pdf.
3 Ibid.
4 Welcome to Government of India | Ministry of Power (powermin.gov.in) accessed on 30 April 2021.
5 Annual Report 2020–2021, Ministry of New and Renewable Energy, Government of India.
6 Executive Summary on Power Sector February 2021, CEA. URL: https://cea.nic.in/wp-content/uploads/executive/2021/03/executive.pdf, accessed on 25 May 2021.

7 https://pib.gov.in/PressReleasePage.aspx?PRID=1681053, accessed on 31 January 2021.
8 https://ieeexplore.ieee.org/document/6450514.
9 www.indiaenvironmentportal.org.in/files/file/State%20Renewable%20Energy %20Action%20Plan%20for%20Assam.pdf.
10 http://cea.nic.in/reports/others/ps/pspa1/large_scale_grid_integ.pdf.
11 www.eqmagpro.com/wp-content/uploads/2017/06/Pump-storage-CEA-2017 -01-24.pdf.
12 https://neepco.co.in/projects/power-potential.

3

CROSS-BORDER ENERGY TRADE AND DEVELOPMENT OF BORDER ECONOMIC ZONES (BEZS) IN NORTHEAST INDIA

Toward an Integrated Regional Program

Prabir De and Venkatachalam Anbumozhi

1 Introduction

India's North Eastern Region (NER) has 9% of India's geographical area and contributes 3% to the country's gross domestic product (GDP). However, given its rich energy resource base and strategic location, the NER has the potential to become a "powerhouse" in terms of renewable energy production and trade. The NER is unique in terms of the economic and industrial development opportunities it offers. About 98% of the region's borders form India's international boundaries; it shares borders with China, Bangladesh, Bhutan, and Myanmar. Given its strategic location, the NER serves as a geographical base for India's growing economic linkages not only with the Association of Southeast Asian Nations (ASEAN) but also with neighboring countries in South Asia, namely, Bangladesh, Bhutan, and Nepal.

Over the past several years, India has been part of a number of regional and subregional initiatives that countries in South and Southeast Asia have taken to deepen their economic integration.[1] Further, as a part of its "Act East" policy, India has increased its engagements with the members of the ASEAN and countries belonging to the East Asia Summit (EAS) or Indo-Pacific.

While the NER has been playing a critical role in fostering these regional economic cooperation initiatives, particularly those with the Southeast and East Asia, regional integration is found to be progressing well in some areas (such as border trade) but slow in some other areas, such as cross-border energy connectivity. Energy sector development could be one of the major means of integrating the NER with the economic centers of South and South East Asian countries (Anbumozhi, Kutani, & Lama, 2019; Singh, 2020). The NER is yet to become a best-case scenario in the backdrop of growing ASEAN–India relations. Therefore, new models of regional partnership are

DOI: 10.4324/9781003433163-3

being experimented with, focusing NER's greater economic engagement with the ASEAN, particularly in the energy sector. The energy sector remains one of the core areas of ASEAN–India relations as noted in the latest ASEAN–India Plan of Action (POA) (2021–2025). Energy sector integration between South and Southeast Asia, using the NER and Myanmar as gateways, has the potential to become a regional power market (USAID, 2016).

While the majority of the population in NER states has electricity grid connection today, supply is often unreliable. Per capita energy consumption in the NER is one of the lowest in the country. A severe shortage of essential energy infrastructure is also undermining efforts to achieve more rapid social and economic development in the NER and Myanmar. Several ASEAN member states and India face common sustainable developmental challenges as well as offering similar commitments in the Paris Climate Agreement, which aim to maximize the use of clean energy sources. Therefore, the ASEAN and India have called for deeper cooperation in the energy sector in general and trade in renewable energy in particular, thereby attaining the Nationally Determined Contributions (NDC) targets and accomplishing Goal 7 of the Sustainable Development Goals (SDGs): affordable and clean energy.

It has been argued that the NER can play a three-dimensional role as power producer, exporter, and transit provider, provided that a quadrangular approach to build energy linkages and promote integration is consciously put in place.[2] The prospect of energy trade between the NER and its immediate neighbors is thus very high and the NER, along with Bhutan and Nepal, could become a subregional energy generation hub (Anbumozhi et al., 2019). On the other hand, however, progress has been limited due to bottlenecks and gaps in energy generation and transmission infrastructure, financial markets, and trade facilitation, as well as trade barriers and limited regional cooperation (Anbumozhi et al., 2019).

Development of mini-grid and off-grid decentralized systems can provide new economic opportunities. The steady and sustainable energy infrastructure linkages can strengthen the NER's trade and economic engagements with the neighboring countries (Brunner, 2010; De et al., 2020; Murayama et al., 2022). Given that positive association between connectivity and trade, a new model of Border Economic Zone (BEZ) is being explored in the country.

The BEZ is defined as a geographically delimited area within which neighboring governments facilitate industrial activity through fiscal and regulatory incentives and energy infrastructure support. The BEZs could attract investments, generate employment, facilitate exports and energy trade, and promote technology and innovation to host countries. The BEZs have gained high popularity on completion of economic corridors in the Greater Mekong Subregion (GMS).[3] Surrounded by an international border, the BEZ could be a natural choice for the NER, which may help it become more economically engaged with Southeast Asia and Bangladesh while promoting

Cross-Border Energy Trade (CBET). This may also reposition the NER to utilize its rich and unharnessed hydro and other modern renewable sources effectively. Moreover, the economic and energy sector reforms in Bangladesh and Myanmar that promote off-grid, decentralized energy supply over the last few years have provided an adequate fillip to the cross-border energy connectivity and energy trade integration in the NER.

A considerable volume of international investments, both in the fossil fuel and renewable energy sectors, in the ASEAN over the last ten years have come from India, which has been already engaged in several small and large-scale infrastructure developments in the region, particularly in Cambodia, Lao PDR, Myanmar, and Viet Nam (many of which are driven by development aid). India has been building hydropower projects, power transmission lines and substations, and oil and gas pipelines in these countries.

To promote people-to-people economic and social connections, the Government of India has been setting up Border Haats across the NER, which shares borders with Bangladesh and Myanmar. Border Haats aim to promote the well-being of the people dwelling in remote areas across the frontiers by establishing a traditional system of marketing the local produce through local markets in local currency and/or barter basis.[4] The following are the operational Border Haats in NER: (i) Kalaichar (Meghalaya–Bangladesh border), (ii) Balat (Meghalaya–Bangladesh border), (iii) Kamlasagar (Tripura–Bangladesh border), and (iv) Srinagar (Tripura–Bangladesh border). In addition, three more Border Haats, namely, Nalikata (India)–Saydabad (Bangladesh); Ryngku (India)–Bagan Bari (Bangladesh), and Bholagunj (India)–Bholagunj (Bangladesh) have been recently inaugurated.[5] Similar initiatives have been proposed in the NER–Myanmar border regions. However, these Border Haats are not meant for attracting large-scale investment including small scale energy infrastructure. To attract investment, particularly FDI in the NER, developing a Border Economic Zone is a promising option, which is being explored in India.

This chapter aims to identify the scope and opportunities in setting up BEZ in the NER, and implications on cross-border energy trade. The study seeks to answer the research questions of what could be done to promote BEZs in India's Northeast and what are the implications for bilateral and international energy cooperation. As the local economy and energy demand grow across the borders, improving energy connectivity is crucial for building greater economic integration between the ASEAN countries and the NER.

This study is an effort to improve existing knowledge on ASEAN–India connectivity and other relevant issues related to energy sector cooperation, and also provide policy direction for strengthening energy market integration between the NER and ASEAN.

The rest of this chapter is organized as follows. Section 2 discusses the concept of BEZ, the relevance for the NER and what it means for energy trade. Section 3 discusses the Mekong subregion's experiences with the BEZs, followed by a discussion on India's experiences with Border Haats in Section 4. Section 5 then discusses possible roadmap to BEZs. Cross-border energy trade and implications for energy cooperation are briefed in Section 6. Section 7 then presents the way forward and conclusions are drawn in Section 8.

2 Concept of Border Economic Zone (BEZ): Stylized Facts

The BEZ, cross-border or otherwise, is identified as a catalyst for economic growth in the border areas.[6] Complementary and differentiated economic resources between the countries sharing a geographical border help facilitate development of border industries, border trade, and border tourism.[7] It is primarily an economic enclave, supported with special incentives and policies to facilitate economic activities and growth of the border areas and the hinterland. It offers a set of differential policies to enhance investment, flows of goods and services, and technology. The Cross-Border Economic Zone (CBEZ) is an extension of BEZ. An efficient CBEZ allows the free flow of raw materials, goods, services, and investments, boosting the supply and value chain and making it more competitive and attractive to foreign investments (Kudo & Ishida, 2013). Several countries have developed BEZs and CBEZs, in which the best example is GMS. For example, the development of Special Economic Zones (SEZs) has generated economic gains in the Lancang–Mekong countries of Cambodia, China, Lao PDR, Myanmar, Thailand, and Viet Nam.[8] Wood and Sizba (2015) identified three options for the location of what they call border development zones (BDZs).

Figure 3.1 illustrates the stages of movability of goods and factors of production and the rise and fall of border industries. When border barriers are very high as in stage 1, we see no scope for growth of the border industry. With a rise in economic integration, scope of border industries grows. In the third stage, economies are connected deeply with the free flow of factors of production, leading to fall of border barriers and industries. The movement from stage 1 to stage 3 generates high value addition to border regions in terms of jobs, production, technology, and trade, among others. Border trade is one of the integrated components of the BEZ.

BEZs have been found to be an effective tool for the development of border areas through exploiting local and cross-border connections. As noted in Wang (2016), a BEZ can strengthen industrial links between the economies on both sides of border, and a CBEZ can generate spillover effects for the economic development of neighboring areas. The BEZ is a popular growth center across many countries, and is sometimes also known as a border industrial park, etc. BEZs are operating successfully at the Viet Nam–Lao PDR border, the China–Viet Nam border, the Thailand–Lao PDR border, the North Korea–South Korea border, and the Viet Nam–Cambodia border, to mention a few.

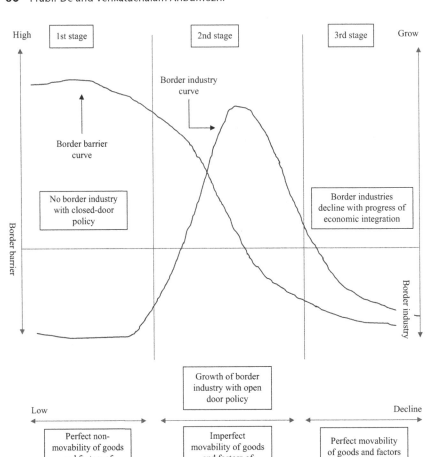

FIGURE 3.1   Movability of goods and factors of production and rise and fall of border industry.

Source: Kudo and Ishida (2013)

Examples of BEZs include the Chaing Rai Border Economic Zone in Northern Thailand, the Savan-Seno Economic Zone at the Thai–Lao PDR border (development of Mukdahan/Savannakhet Border), the Lao Bao Border Free Trade Zone at the Lao PDR–Viet Nam border along GMS economic corridor, the Ruilli Border Economic Cooperation Zone between China and Myanmar, and the Moc Bai Cross-Border Economic Area between Viet Nam and Cambodia. Looking at the success of BEZs, China and Viet Nam have planned two BEZs at border crossing points: Hekou–Lao Kai and Pingxiang–Dong Dang.

Figure 3.2 presents the illustration of the CBEZ concept in the context of India's North East. In India, Border Haats have been set up in India's North

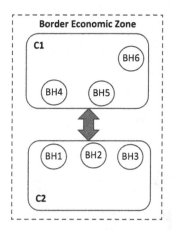

FIGURE 3.2   Illustration of Cross-Border Economic Zone in context of India
Note: C1, C2: Country 1 and Country 2. BH1–BH6: Border Haat 1
to Border Haat 6.

Source: Author

Eastern Region along the India–Bangladesh border to foster border connectivity. India has set up four Border Haats along the India–Bangladesh border.

While the Border Haat is a successful project, it cannot generate market-driven larger exchange of goods and services across borders due mainly to space limitations and lack of infrastructure. Elevating Border Haats into CBEZs may help the North Eastern states to benefit from a scale economy. Social and economic benefits would be much larger in the case of BEZs, subject to certain conditions. Not only would BEZs offer more space for industries, they would also facilitate the 'Make-in-India' initiative in the NER. Several cross-border connectivity projects in the NER such as the Trilateral Highway, India–Bangladesh border connectivity programs, exchange of electricity between India and Myanmar and India and Bangladesh, etc., have gained momentum in recent years.

Figure 3.3 presents a schematic illustration of a CBEZ. Given its locational advantage, setting up BEZs may help the NER to build the needed infrastructure and trade facilitation services, thereby facilitating economic linkages between the countries. In particular, BEZs will promote energy trade in the NER and strengthen economic links with Myanmar and Bangladesh.

3 BEZs in Mekong Subregion and Lessons They Offer

The Mekong subregion is known for the world's most successful BEZs. It has become a symbol of the intensive border exchange between countries. Since the early 2000, BEZs in the Mekong subregion have attracted a great amount of foreign investment. These BEZs offer some important lessons to enhance border connectivity and regional integration in the Northeastern

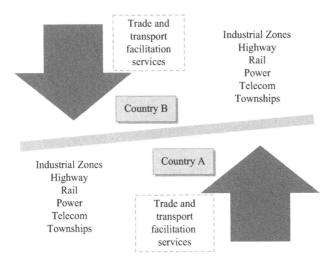

FIGURE 3.3 Illustration of a CBEZ.

Source: Author

part of India. To date, the Mekong subregion has over two and a half dozen economic zones, including 15 BEZs.[9] Table 3.1 and Map 1 present the list of BEZs in the Mekong subregion.

The Asian Development Bank (ADB) has promoted the BEZs along GMS economic corridors since 1998. The establishment of economic corridors added new momentum to the development of BEZs. Later, the number of BEZs has grown from fewer than 6 in the early 2000s to over 18 in 2019. The ADB has funded development of three economic corridors in Mekong subregion, namely, the Southern Economic Corridor, the North–South Economic Corridor, and the East–West Economic Corridor. Later, four more economic corridors were added, and many of the BEZs are located along the corridors.[10]

The GMS BEZs offer the following lessons for the development of BEZs in India's NER.[11]

- Some of the key success factors of BEZs include (i) provision of modern infrastructure, advanced border crossings, and cheap and reliable utilities such as electricity, water, telecommunications, and waste disposal; (ii) free movement of goods and people, including visa-free movement or visa on arrival, and discretionary quotas for work permits; (iii) establishment of dedicated customs facilities, testing and certification labs for agriculture and livestock, warehouse facilities; (iv) relaxing rules of origin for goods processed in BEZs; (v) quality logistics, and specifically road and sea linkages along the corridor; (vi) independent SEZ governing authorities such as one-stop service centers; (vii) transparent standards and consistent policies with strong support from local and national government; and (viii) identification of industrial clusters.[12]

TABLE 3.1 List of Border Economic Zones and Economic Corridors

Sr. No.	Routes	Border
1	Southern Economic Corridor	• Aranya Prathet (Thailand)–Poipet (Cambodia) • Bavet (Cambodia)–Moc Bai (Viet Nam) • Cham Yeam (Cambodia)–Hat Lek (Thailand)
2	East–West Economic Corridor	• Myawaddy (Myanmar)–Mae Sot (Thailand) • Savannakhet (Lao PDR)–Mukdahan (Thailand) • Lao Bao (Viet Nam)–Dansavanh (Lao PDR)
3	North–South Economic Corridor	• Boten (Lao PDR)–Mohan (China) • Chiang Khong (Thailand)–Houayxay (Lao PDR) • Mae Sai (Thailand)–Tachilek (Myanmar) • Hekou (China)–Lao Cai (Viet Nam)
4	Other routes	• Ruili (China)–Muse (Myanmar) • Nongnokkhien (Lao PDR)–Trapeang Kreal (Cambodia) • Thanaleng (Lao PDR)–Nong Khai (Thailand) • Namphao (Lao PDR)–Cau Treo (Viet Nam) • Vangtao (Lao PDR)–Chong Mek (Thailand)

Source: ADB (2018)

- Proper coordination between BEZ authorities on both sides of the border is essential for success of CBEZs. At the country level, governments must cooperate closely in order to make CBEZs successful.
- Regulatory aspects such as laws, regulations, procedures, etc., and development plans, trade-related policies, etc., need to be harmonized between participating countries in BEZs.
- All round border area development is needed for a comprehensive development of the bordering economies and regional integration.
- Local businesses and communities must be included in collaborative development of BEZs.

What follows is that the success of BEZs appears to be related to the infrastructure they provide (transport, reliable power supply, etc.) and streamlining of regulations, rather than the tax and other financial incentives used to promote early SEZs.

Several case studies in Myanmar in the East–West Economic Corridor clearly show that electricity provision and energy trade, both non-conventional and grid-connected, improve quality of life significantly. Some of the positive impacts of such energy connectivity include wealth generation, communication, trade, health, education, income, and employment (Yoshikawa and Anbumozhi, 2017).

4 Development of Border Haats in Northeast India and the Way toward BEZs

To foster border connectivity, Border Haats have been set up in the NER, particularly along India–Bangladesh border. These Haats are fenced areas

MAP 3.1 BEZs in Mekong Subregion (a) Agartala–Chattogram BEZ (b) Shillong–Sylhet BEZ (c) Moreh–Tamu BEZ

Source: ADB (2018), available at www.adb.org/sites/default/files/institutional-document/470781 /role-sezs-gms-economic-corridors.pdf

equipped with market sheds, and me dical and security booths. India presently has four Border Haats in operation along India–Bangladesh border, of which two (Srinagar and Kamalasagar) are located in Tripura state and other two (Kalaichar and Balat) in Meghalaya state (Map 2). Besides, two Border Haats in Tripura at Palbas and Kamlapur and four in Meghalaya at

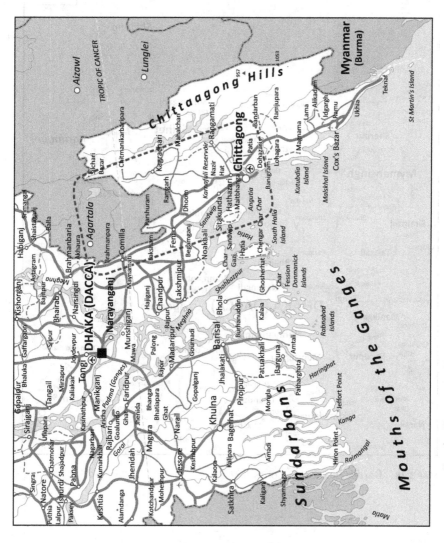

MAP 3.1A Border Economic Zones

MAP 3.1B Border Economic Zones

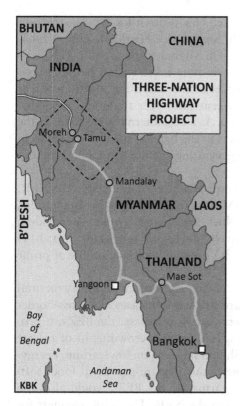

MAP 3.1C Border Economic Zones

Bholaganj, Nalikata, Shibbari, and Ryngku have been agreed upon by the two countries.

The rules and regulations of trade in Border Haats require sellers and buyers to reside within a five-kilometer radius of a particular Border Haat. No duties are imposed on trading at Border Haats, and both Indian Rupee and Bangladeshi Taka are accepted in these markets. In a way, trade takes place at a local currency level.[13] However, the facilities are inadequate to cater to the local trade. The traded items are mostly household items, horticultural products, processed foods, clothes, spices, vegetables, etc. Border Haats open at particular times during the day and operate on a fixed weekly schedule.

India and Bangladesh have agreed to set up a new Border Haat at the Sonamura subdivision of the Sepahijala district of Tripura to boost trade and commerce between the two countries. In October 2019, the Chief Minister of Tripura announced that two more Border Haats have been approved for launch at Dharmanagar and Kamalpur in the North and Dhalai districts of Tripura.

The success of the Border Haats with Bangladesh has prompted other North Eastern states of India to pursue similar arrangements with Myanmar. At the request of the Mizoram government, the center is likely to set up four Border Haats along the border with Myanmar to boost local trade and the livelihood of the people living there. The location for the construction of Border Haats has been identified in four places: Hnahlan, Zote, Vaphai (Saikhumphai), and Sangau (Pangkhua) in Mizoram.

To conclude, Border Haats have been quite successful in building people-to-people links and improving border connectivity. But, due to inherent constraints, they may not be able to generate larger trade and economic development for the adjacent hinterland. As a best alternative, opting for the path of BEZ gradually with due address to the barriers mentioned may help the Northeastern states not only in generating further economic activities but also achieving greater welfare.

Based on location, three BEZs are proposed to be developed in the NER. They are between Manipur and the Myanmar border (Moreh–Tamu area), Meghalaya–Bangladesh (Shillong–Sylhet area), and Tripura–Bangladesh (Agartala–Chittagong area), respectively. Table 3.2 and Maps 3 (a, b, c) illustrate these BEZs, whereas Table 3.3 presents the basic profile of prime hubs of the BEZs.

While these BEZs are well-connected and endowed with infrastructural facilities, the cross-border linkages are still weak in cases like cross-border energy linkages or transportation services. Nonetheless, the BEZs, if promoted, may attract new investments, leading to a crowding-in of private sector investments. However, focus should be given on low-carbon, energy-efficient industries. Some of the target industries are processed food, software, electronics, education, health, garments, etc. BEZs would also aim to develop smart industries and cities in the NER. The NER accounts for only a very small share of cumulative historical energy-related carbon emissions. As BEZ-driven economic growth continues to move in the direction of energy-intensive economic activity, future developments should be consistent with energy security and sustainable development goals.

5 Trade, Local Industries, and Energy Strategy to Unlock the BEZ's Potential

Border trade is one of the integrated components of the BEZ. The NER contributes only 8 to 10% of India's export to Bangladesh. Table 3.4 presents

TABLE 3.2 Proposed BEZs in NER

Border Zone	Location
Border Economic Zone 1	Agartala (Tripura, India)–Chattogram (Chittagong Division, Bangladesh)
Border Economic Zone 2	Shillong (Meghalaya, India)–Sylhet (Sylhet Division, Bangladesh)
Border Economic Zone 3	Moreh (Manipur, India)–Tamu (Sagaing Region, Myanmar)

TABLE 3.3 Basic profile of BEZs hubs (as of 2019)

	Shillong (India)	Sylhet (Bangladesh)	Agartala (India)	Chittagong (Bangladesh)	Moreh (India)	Tamu (India)
Population*	143,000	526,000	438,000	2,600,000	16,847	68,603
Area**	22,429	12,298	10,492	5,283	12	28
Infrastructure						
Gas	No	Yes	Yes	Yes	No	No
Rail	Yes	Yes	Yes	Yes	No	No
Highway	Yes	Yes	Yes	Yes	Yes	Yes
Air	Yes	Yes	Yes	Yes	No	No
Sea Port	No	No	No	No	No	No
Land port	Yes	Yes	Yes	Yes	Yes	Yes
Digital access	Yes	Yes	Yes	Yes	No	No
Warehouses	Yes	Yes	Yes	Yes	No	No
Electricity	Yes	Yes	Yes	Yes	Yes	Yes
Banks	Yes	Yes	Yes	Yes	Yes	Yes
Townships	Yes	Yes	Yes	Yes	No	No
Cross-border linkages						
Trade	Yes	Yes	Yes	Yes	Yes	Yes
Immigration	Yes	Yes	Yes	Yes	Yes	Yes
Electricity grid linkages	No	No	Yes	No	No	No
Gas linkages	No	No	No	No	No	No
Bus services	No	No	Yes	No	No	No
Railway linkages	No	No	No	No	No	No
Road linkages	Yes	No	Yes	No	Yes	Yes

(Continued)

TABLE 3.3 (Continued)

	Shillong (India)	Sylhet (Bangladesh)	Agartala (India)	Chittagong (Bangladesh)	Moreh (India)	Tamu (India)
Digital linkages	No	No	No	No	No	No
Banks	Yes	Yes	Yes	Yes	Yes	Yes
Tourism	Yes	Yes	Yes	Yes	Yes	Yes
Major industries	Agriculture and food processing, floriculture, horticulture, mining, cement, tourism, hydroelectric power, handlooms, handicrafts and sericulture	Tea, cement, food processing, fertilizer, ceramics, machinery and equipment, ready-made garments and pharma	Agriculture, rubber, tea, handicraft, rubber	Tea, petroleum, oil refinery, chemicals, pharma, food processing	Handicrafts, processed food, agriculture, plastics	Agriculture, processed food, handicrafts
Tariff concessions+	Yes	Yes	Yes	Yes	Yes	Yes
FTA benefit++	Yes	Yes	Yes	Yes	Yes	Yes

Source: Author's own based on several secondary sources

*As of December 2019 **Tentative +Preferential concessions such as GSP, GSP+, DFQF, etc. ++Preferential or free trade agreement or both

TABLE 3.4 Trends in NER's trade with Bangladesh

Year	NER's export to Bangladesh	NER's import from Bangladesh	NER's total trade with Bangladesh
	(US$ million)		
2010–2011	91.56	64.51	156.07
2011–2012	134.62	81.94	216.56
2012–2013	171.23	75.77	247.00
2013–2014	172.71	80.89	253.60
2014–2015	221.30	85.54	306.84
2015–2016	229.39	83.88	313.27
2016–2017	251.54	87.18	338.72
2017–2018	269.07	112.50	381.57
2018–2019	280.44	145.32	425.76
2019–2020	297.10	132.90	430.00

Source: Authors, based on Indian customs

TABLE 3.5 Trading with Myanmar: Number of LCSs

NER State	LCS in India	LCS in Myanmar
Arunachal Pradesh	Nampong* (Pangsau Pass)	Pangsu
Manipur	Moreh	Tamu
Mizoram	Zokhawthar (Champai)	Rih**
Nagaland	Avangkhu***	Somara

Source: De (2013) based on various sources

the trends in the NER's export to and import from Bangladesh. The NER's export to Bangladesh has increased in recent years, whereas import is rising at a slower pace. The NER has a trade surplus with Bangladesh. The total trade between the NER and Bangladesh is US$ 430 million in 2019–2020, which has increased from US$ 156.07 million in 2010–2011. The NER's exports to Bangladesh are mostly primary horticulture items and minerals such as coal, quick lime, ginger, oranges, dry fish and other citrus fruits, boulder stones, dry fish, raw hides, woven fabrics and synthetic filament, etc. On the other hand, NER's imports from Bangladesh are well diversified and mostly secondary items such as cement, processed foods, plastics, knitted and crocheted synthetic fabrics, garments, cement, fish, PVC pipes, wooden furniture, etc.

Since 2015, trade at the border between Myanmar and India has been conducted based on a Most Favoured Nation (MFN) basis. Although the NER's trade with Myanmar at the border has been rising fast, the NER's informal trade volume with Myanmar at the border has been more than its formal trade. Table 3.5 presents a number of Land Custom Stations (LCSs) dealing with border trade with Myanmar. All four NER states have functional LCSs with Myanmar, of which Moreh LCS (Manipur) is the biggest

TABLE 3.6 Trends in NER's trade with Myanmar at More–Tamu Border

Year	NER's export to Myanmar	NER's import from Myanmar	NER's total trade with Myanmar
	(US$ million)		
2010–2011	4.50	8.30	12.80
2011–2012	6.54	8.87	15.41
2012–2013	11.67	26.96	38.63
2013–2014	17.71	30.92	48.63
2014–2015	18.11	42.61	60.72
2015–2016	18.62	53.02	71.64
2016–2017	20.65	97.22	117.87
2017–2018	21.87	143.90	165.77
2018–2019	23.45	177.20	200.65
2019–2020	14.10	120.16	134.26

Source: Authors, based on Indian customs

one in terms of volume of trade, with the Zokhawthar/Champai LCS (in Mizoram) coming second.

NER's export to Myanmar through land-border has increased from US$ 4.50 million in 2010–2011 to US$ 23.45 million in 2018–2019 and then declined to US$ 14.10 million (Table 3.6). NER's import from Myanmar was almost four times more than its export in 2018–2019. In 2018–2019, NER's import from Myanmar has increased to US$ 177.20 million from US$ 8.30 million in 2010–2011, and then declined to US$ 120.16 million in 2019–2020. The total trade with Myanmar stands at US$ 134.26 million in 2019–2020. In 2019–2020, NER's exports to Myanmar were cumin seed, cotton yarn, auto parts, soya bean meal, wheat flour, and pharmaceuticals, whereas imports were betel nuts, dry ginger, green mung beans, black matpe beans, turmeric roots, resin, and medicinal herbs. However, informal trade between the NER and Myanmar has been carried out extensively, and some of the Indian products traded between them through informal channels are food items, cosmetics, wood and timber products, construction materials, garments, gas cylinder, medicines, plastics and materials, rubber products, solar and electrical items, betel nuts, auto parts, petroleum products, etc.

5.1 Local Industries in Manipur[14]

5.1.1 Tera-Urak Industrial Estate

There are 27 units that have taken manufacturing sheds, and 11 units are running at the Tera-Urak Industrial Estate.

(i) Kitawan Unit

- Manufactures garments based on power looms such as fabric, tailoring, embroidery, etc.
- Annual sales volume of Rs. 15 million.
- Has about 15 people working.
- Raw materials are exported from Kolkata.
- Most of the challenges are transportation of raw materials, logistics, availability of electricity, working capital.

(ii) Papad Unit

- Has about 34 people working in two shifts. Currently, the unit is not working because of a breakdown.
- Manufactures food items such as papad.
- Started working from November 2018 with an investment of Rs. 4 million.
- Pays Rs. 6,000 per month rent to Manipur government
- The main problems are electricity and water.

(iii) Sangai Water Tank Unit

- The Sangai Water Tank unit manufactures 1,000 liter water tanks.
- At the moment, five people are working.
- It manufactures 24 water tanks per day (600 liter capacity) and 12 water tanks per day (1,000 liter capacity).
- The main problem is the electricity and water.

(iv) PVC Pipe Manufacturing Unit

- This unit started operating from November 2016.
- Manufactures PVC pipes of different types, including PVC pipes based on recycled materials.
- Raw materials are imported through Kolkata.
- Transportation is the main hurdle. Rs. 150,000 per truck is the transportation cost for bringing raw materials from Kolkata and Rs. 1.15 million per truck transportation cost for bringing the raw materials from Delhi – first by rail to Guwahati and then by road to Manipur.
- Problems are related to logistics, working capital and electricity.

(v) Electrical Transformer Manufacturing Unit

- This unit is yet to be operational. It is in place but has not yet started manufacturing electrical transformers.

- Initial investment has already been made.
- Technology is sourced internally and standard of BIS has been followed.
- At the moment, 25 people are working.

5.1.2 Takyel Industrial Estate

There are 35 units working at present, and most of the units are plastics, handlooms, processed foods etc.

(i) Manipur Creations

- The company has been started operating since 2012 in Tirupur (Tamil Nadu). In 2015, production shifted to Imphal.
- Current sales volume is Rs. 6.7 million per month. The business model is based on self-help groups. Presently, it has 100 self-help groups trained by the company and scattered around Imphal city. There are 146 people working with the Manipur Creations.
- Financial support toward the training of households is sourced from TATA Trust.
- Manipur Creations would like to export to Myanmar if opportunity arises.
- Major challenge for expansion of business is uncertainty in delivery of raw materials and finished goods.

(ii) KSHIS Plastics

- The company has started operating in 1994. It has been working for the last 25 years. This is a Micro, Small and Medium Enterprise (MSME).
- At the moment, 54 people are working with the company but all are contractual employees.
- It manufactures plastic bags, pet bottles for water, juices, plastic caps etc. Initially, the company owner invested Rs. 50,000.
- The company has an annual turnover of Rs. 500,000 with monthly sales of Rs. 450,000–500,000.
- Most of the products are locally consumed.
- Company promoters think that they cannot export because they would not be able to compete with the other domestic manufacturers.
- When this industrial estate started working, there were 3–4 plastic units. Except for KSHIS Plastics, all other companies had closed down.
- If the government stops importing through Moreh, KSHIS can export to Myanmar and other countries. Illegal import is killing the business.
- The company has already received industrial incentives from the Government of India and Government of Manipur, including a ten-year tax holiday.

- The major constraints are the physical space in the industrial estate (the company cannot expand), insurgency, informal tax, etc.
- Specific recommendations: (i) government support is needed for capacity building of human resources; (ii) stop illegal market and import from neighboring countries; and (iii) reduce the current GST rate of 18%.

5.1.3 Food Park in Nila Kothi, Imphal

(i) Shri Balaji Flour Mill

- It exports Maida to Myanmar since 2017.
- Every year 5,000 to 6,000 packets of Maida are exported from this factory (each packet is 45 kg).
- Export is going via the Moreh border. In 2018–2019, about 2,000 packets were exported. Price was Rs. 1,380 per packet.
- In total, 50 people are now working.
- Domestic Production:
  - Maida – 128,000 packets produced in 2018–2019.
  - Atta – 61,000 packets produced in 2018–2019.
  - Suji (Rawa) – 2,900 packets produced in 2018–2019.
  - Atta with Bran – 8,300 packets (each packet – 30 kg) produced in 2018–2019
  - The mill also produces Mattar Dal, Besan, Soyabean, etc.
- The major problems relate to transportation and logistics. Most of its agricultural raw materials are sourced from Punjab. Raw materials are transported to Dimapur by rail and from Dimapur to Imphal by road. The road transportation from Dimapur to Imphal faces insurgency.

(ii) Likla Industries

- It produces drinking water, fruit drinks, bread, etc.
- Annual sales volume is about Rs. 300 million per annum.
- Company exports to Myanmar via third party.
- The company has two business units at this food park. It is coming up with one more unit for which the construction is going on and is likely to be completed by 2019.
- The main problem that the company has been facing is law and order.
- There should be free movement of goods between Moreh and Dimapur and Moreh and Silchar. Infrastructure in Moreh has to be improved. Also reduce the GST, which is presently very high – 12% on food items and 18% on processed food items and 18% on plastics. No exemption has been given to Likla Industries.
- To date, the company has received a 15% subsidy on plant and machinery and this should be increased to 30%.

- Electricity condition has improved.
- Specific recommendations: the Ministry of Food Processing should provide a construction subsidy. Sales Tax concession should be given to the new unit for ten years.

5.2 Strategy to Unlock the BEZ's Potential

We have to draw on local advantages, e.g., low-wage or labor-intensive activities, to become competitive. Myanmar has received the GSP benefits, similar to Cambodia or Lao PDR in Mekong. India (and South Asian countries) may benefit if BEZs are promoted, particularly with Myanmar. BEZs can also be developed along the India–Bangladesh border. We have to classify the economic activities such as border trade, border industries, border tourism, etc. Target industries could be horticulture, processed food, software, electronics, education, health, garments, consumer goods, wood products, handicrafts, etc., BEZs can provide additional support to cross-border value chains.[15]

The border trade infrastructure at the NER is still inadequate to support the rising trade volume. The NER needs drastic improvement in border infrastructure, particularly when dealing with trade with Bangladesh. The benefits of connectivity corridors will flow only when border infrastructure is upgraded to facilitate trade and investment at the border region.

The local industries have high business prospects provided there is infrastructure at the industrial estates, security and safety, and access to international markets. Informal trade coupled with insurgency have de-motivated the local industries to grow. The BEZ may promote local industries with scale and border benefits.

India may consider building BEZs across the India–Myanmar (and also the India–Bangladesh) border. As outlined in the previous section, Mekong countries offer rich lessons. Similar to GMS economic corridors, several cross-border connectivity corridors are under construction such as the Trilateral Highway, Kaladan Multimodal Transit Transport Project, etc. A possible extension of the India–Myanmar–Thailand Trilateral Highway to Cambodia, Lao PDR, and Viet Nam is also under consideration. The protocol of the India–Myanmar–Thailand Motor Vehicle Agreement (Trilateral MVA) is being negotiated. This agreement will have a critical role in realizing seamless movement of passenger, personal, and cargo vehicles along roads linking India, Myanmar, and Thailand.

In North East India, we find that the border barrier has been declining. To give a big push to border connectivity, we need to facilitate industries in the border region.[16] Once this happens, firms will be motivated to locate the factories in the border region. The NER can enter into the second stage (as illustrated in Figure 3.1). So, it is the right moment to promote the development of border areas as border economic zones in NER.

5.3 Strategy for Aligning Energy Development with Local Industrial Transformation

To infuse greater economic benefits at the local level, the stability and reliability of the power system is a must (Anbumozhi et al., 2019). In the backdrop of growing connectivity linkages between the NER and the ASEAN, cheaper and quality electricity coupled with faster movement of goods and services may lead to local industrial transformation in the NER. Savings in terms of time and costs resulting from cheaper electricity and connectivity may translate into more disposable income, particularly among MSMEs. Moreover, as noted in the Economic and Social Commission for Asia and the Pacific (UNESCAP, 2018), energy connectivity enables neighboring countries to integrate their markets, which may incentivize investment in domestic power-generation facilities and thereby contribute to lower electricity prices in the long term. However, the occurrence and extent of these benefits may differ significantly depending on several micro and macro issues such as grid inter-connectivity, regulatory convergence, price, technology, among others. A BEZ can offer benefits of dual factors: one due to pull/demand factor (e.g. energy exchange, regulations, etc.) and other due to push/supply factor (e.g. production, technology, grid inter-connectivity, etc.).

It goes without saying that local industrial development is about to happen in the NER. To enhance such development, the NER has to design an appropriate strategy to unlock the region's rich renewable energy potentials such as hydro, solar, and wind energy along with cross-border energy engagements through energy pipelines and grid inter-connectivity. A successful program for a grid-based renewable in the NER is stimulating private investment to have the best mix of both types of energy, currently dominated by coal. Expectations are also high in several NER states with recent gas discoveries, which can spur further industrial development. National priorities are equally important.[17] For example, significant emphasis has to be given to the transmission network within the NER states, and also with the rest of India. The central government and those of the statesare also determined to pursue the challenging and long-term endeavor of promoting gas use in domestic power generation and industry along with an increase in a mini-grid-based solar system. Here, governments have a strong role in policy formulations for prioritized investment needs, including regulatory reforms, financial flows, and institutional cooperation at national and international levels (Anbumozhi, 2019). Developing a local industrial base for new energy demand and supply requires a careful choice and location of projects to anchor the development of distributed energy systems, in an appropriate way to get economic and environmental value from both renewable and other conventional energy sources.

6 Cross-Border Energy Trade and BEZs

The electricity sector leads energy market integration (Wu et al., 2012). With a rise in economy, energy trade is likely to pick up momentum between India and its South Asian neighbors. Presently, India exports electricity to Nepal, Bangladesh, and Myanmar, while India imports power from Bhutan. However, sometimes India also exports power to Bhutan during lean hydro season. Import or export of energy by India into/from these countries during the Financial Year 2019–2020 (as of January, 2020) is given in Table 3.7.

India's Ministry of Power (MoP) has notified the Guidelines on Cross-Border Trade of Electricity. The key objective is to facilitate and promote cross-border electricity trade, and ensure greater transparency, consistency, and predictability in regulatory approaches across jurisdictions while maintaining grid security.

India has signed the Memorandum of Understanding (MoU) with Bhutan, Bangladesh, Nepal, and Myanmar to, inter alia, improve power connectivity with these neighboring countries. The Indian Ministry of Power issued the Guidelines on Cross-Border Trade of Electricity on 5 December 2016, which was subsequently substituted by the Guidelines for Import/Export (Cross-Border) of Electricity–2018, issued on 18 December 2018, to promote cross-border trade of electricity with neighboring countries. The Central Electricity Regulatory Commission issued CERC (Cross-Border Trade of Electricity) Regulations, 2019 on 8 March 2019. To further improve power connectivity with neighboring countries, the following interconnections are at various stages of implementation:

- 400kV operation of Muzaffarpur (India)–Dhalkebar (Nepal) 400kV D/c line (operated at 220kV).
- Baharampur (India)–Bheramara (Bangladesh) 2nd 400kV D/c line.
- Alipurduar (India)–Jigmeling (Bhutan) 400kV D/c (Quad) line.
- Gorakhpur (India)–New Butwal (Nepal) 400kV D/c (Quad) line.
- Sitamarhi (India)–Dhalkebar (Nepal)–Arun-3 HEP (Nepal) 400kV D/c (Quad) line.

TABLE 3.7 India's trade in electricity with immediate neighbors (as of January 2020)

Sl. No.	Country	Import (Million Units)	Export (Million Units)
1	Nepal	-	1839.25
2	Bhutan	6165.78	-
3	Bangladesh	-	6168.14
4	Myanmar	-	7.34

Source: PIB

Nepal and Bangladesh are not using the Indian transmission grid to trade power between them. The trading arrangement for import/export of electricity with neighboring countries, including Nepal, Bhutan, and Bangladesh, would facilitate regional power trade and help in meeting the energy demand in the respective countries, thereby moving toward greater energy security in the region. This would also facilitate the development of BEZs. However, it is essential to understand that the institutional structure and the stages of evolution of regulatory regimes in countries' power sector impacting cross-border energy trade is different. The bulk of energy generation assets in these countries remain state-owned, and many of the Independent Power Producers (IPP) are made up of small-scale, costly plants that started operations in response to economic incentives for renewable energy integration.

India already has regional power system integration with Bangladesh, Bhutan, and Nepal through synchronous (High Voltage Alternating Current, HVAC) and asynchronous (High Voltage Direct Current, HVDC) connections. Learning from the best practices of advanced nations, the latest technologies like STATCOM, Voltage Source Converter-based HVDC system, etc., have been deployed in the Indian grid as a continuous measure of improvement for facilitating power transfer with reliability amongst regional neighboring countries.

6.1 The NER's Energy Supply and Facilitation of BEZ

In the previous sections, we have identified the location of three BEZs, of which two are with Bangladesh and one with Myanmar. Uninterrupted supply of electricity is needed for the success of the BEZs. Use of renewable energy is also crucial for the development of BEZs. At the same time, the supply of energy in the NER is a big challenge. Table 3.8 presents an actual power supply scenario in the NER. As of 2018–2019, except for the state of Sikkim, the rest of the NER is shown as having an energy deficit, and will continue to be an energy-deficit region (Table 3.9, which presents the NER's electricity demand forecast for 2021–2022 and 2026–2027). The NER's installed power generation capacity for various states is given in the Appendix (Table A3.1). Table A3.2 in the Appendix presents power projects which are currently under operation in the NER. Most of the power projects in the NER are hydro plants, but there are some thermal coal- and oil-based power plants in Assam. In the coming years, the NER will depend more on renewable energy. In terms of hydropower, the NER has the potential of about 58,971 MW (almost 40% of the country's total hydro potential).[18] In addition, the NER also has an abundant resource of coal, oil and gas for thermal power generation. According to the NEEPCO, the NER is blessed with huge hydro potential of about 58,971 MW, out of which 1,727 MW (about 2.92%) has so far been harnessed as on 1 July, 2020. An additional 2,300

TABLE 3.8 Actual Power Supply in NER, 2018–2019

State/Region	Requirement (MW)	Availability (MW)	Surplus (+)/ Deficit (−) (MW)	Peak Demand (MW)	Peak Met (MW)	Surplus (+)/ Deficit (−) (MW)
Arunachal	36,211.23	35,794.53	−416.7	150	148	−2
Assam	398,615.2	384,947.5	−13,667.8	1,865	1,809	−56
Manipur	37,711.35	37,294.65	−416.7	219	216	−3
Meghalaya	81,548.19	81,506.52	−41.67	374	372	−1
Mizoram	26,793.81	26,460.45	−333.36	121	119	−2
Nagaland	37,002.96	33,127.65	−3,875.31	156	138	−18
Tripura	77,631.21	76,714.47	−916.74	298	293	−5
Sikkim	21,960.09	21,960.09	0	106	106	0
NER	717,474.1	697,847.5	−19,626.6	3073	2,956	−117
All India	53,112,374	52,817,808	−7,070	17,7022	17,5528	−1,494

Source: CEA (2020)

TABLE 3.9 NER electricity demand forecast 2021–2022 and 2026–2027

Region	Electrical Energy Requirement (MW)		Peak Electricity Demand (MW)	
	2021–2022	2026–2027	2021–2022	2026–2027
Arunachal Pradesh	62,421.66	108,383.7	278	482
Assam	585,505.2	852,651.5	2,713	4,166
Manipur	87,632.01	137,511	410	667
Meghalaya	106,925.2	132,385.6	488	605
Mizoram	36,086.22	54,462.69	171	252
Nagaland	47,045.43	63,505.08	234	322
Tripura	66,463.65	80,423.1	391	495
Sikkim	26,585.46	33,752.7	170	216
NER	101,8706	146,3075	4,669	6,926
All-India	65,256,178	85,316,575	225,751	298,774

Source: CEA (2018)

MW of hydropower are under construction. The balance of about 93.17% is yet to be exploited. Besides, the NER has 195.68 billion cubic meters (BCM) of gas reserves against 1,380.63 BCM of reserves in the country.

In India, renewable energy may account for 55% of the total installed power capacity by 2030. In the renewable sector, the NER's target is 1,779 MW by 2022 (Table 3.10). The current capacity as of April 2020 is only 417 MW and the balance is to be added in the next 2 years. Therefore, 1,362 MW will be added by 2022 in the renewable energy sector in NER. Most of it is demanded in Assam, followed by Meghalaya, Tripura, and other NER states. Renewable energy has the potential to create employment opportunities at all levels. The best option would be to encourage the private sector in renewable energy production. The NER states must show interest in an increased use of clean energy sources.

India has also started supplying power to Myanmar across the Moreh–Tamu border in April 2016 following Myanmar's request during the first meeting of the India–Myanmar Joint Consultative Commission in July 2015. Subsequently, in October 2016, the two countries signed an MoU on power and are considering strengthening the transmission network to increase power supply in the future. However, with the establishment of a BEZ at Moreh–Tamu, electricity exchange between the two countries will eventually go up.

At the moment, unlike Moreh–Tamu, Agartala and Chittagong or Shillong and Sylhet do not have a formal energy exchange, and their grids are not interconnected. However, if the regional electricity market comes into effect, cross-border electricity exchange will become smooth and

TABLE A3.1 Install capacity (MW) of power utilizes in the NER (as of 30 September 2020)

| State | Ownership/ Sector | Mode wise Break-up | | | | | Nuclear | Hydro | RES * | Grand Total |
| | | Thermal | | | | | | | (MNRE) | |
		Coal	Lignite	Gas	Diesel	Total				
Arunachal Pradesh	State	0	0	0	0	0	0	0	107.11	107.11
	Private	0	0	0	0	0	0	0	29.61	29.61
	Central	37.05	0	46.82	0	83.87	0	330.55	0	414.42
	Sub-Total	37.05	0	46.82	0	83.87	0	330.55	136.72	551.14
Assam	State	0	0	329.36	0	329.36	0	100	5.01	434.37
	Private	0	0	24.5	0	24.5	0	0	48.45	72.95
	Central	403.5	0	435.56	0	839.06	0	389.58	25	1,253.64
	Sub-Total	403.5	0	789.42	0	1,192.92	0	489.58	78.46	1,760.96
Manipur	State	0	0	0	36	36	0	0	5.45	41.45
	Private	0	0	0	0	0	0	0	6.36	6.36
	Central	47.1	0	71.57	0	118.67	0	95.34	0	214.01
	Sub-Total	47.1	0	71.57	36	154.67	0	95.34	11.81	261.82
Meghalaya	State	0	0	0	0	0	0	322	32.53	354.53
	Private	0	0	0	0	0	0	0	13.92	13.92
	Central	50.62	0	109.69	0	160.31	0	79.77	0	240.08
	Sub-Total	50.62	0	109.69	0	160.31	0	401.77	46.45	608.53
Mizoram	State	0	0	0	0	0	0	0	36.47	36.47
	Private	0	0	0	0	0	0	0	1.53	1.53
	Central	31.05	0	40.46	0	71.51	0	97.94	0	169.45
	Sub-Total	31.05	0	40.46	0	71.51	0	97.94	38	207.45
Nagaland	State	0	0	0	0	0	0	0	30.67	30.67
	Private	0	0	0	0	0	0	0	1	1
	Central	32.1	0	48.93	0	81.03	0	61.83	0	142.86
	Sub-Total	32.1	0	48.93	0	81.03	0	61.83	31.67	174.53

Sikkim	State	0	0	0	0	0	0	360	52.11	412.11
	Private	0	0	0	0	0	0	96	0.07	96.07
	Central	102.25	0	0	0	102.25	0	64	0	166.25
	Sub-Total	102.25	0	0	0	102.25	0	520	52.18	674.43
Tripura	State	0	0	169.5	0	169.5	0	0	16.01	185.51
	Private	0	0	0	0	0	0	0	4.41	4.41
	Central	56.1	0	436.95	0	493.05	0	68.49	5	566.54
	Sub-Total	56.1	0	606.45	0	662.55	0	68.49	25.42	756.46
NER	State	0	0	498.86	36	534.86	0	782	285.36	1,602.22
	Private	0	0	24.5	0	24.5	0	96	105.35	225.85
	Central	759.77	0	1,189.98	0	1,949.75	0	1,187.5	30	3,167.25
	Sub-Total	759.77	0	1,713.34	36	2,509.11	0	2,065.5	420.71	4,995.32
All India	State	65,631.5	1,290	7,119.86	236.01	74,277.4	0	26,958.5	2,381.03	103,617
	Private	74,173	1,830	10,598.7	273.7	86,875.5	0	3,394	85,216.1	175,486
	Central	59,790	3,140	7,237.91	0	70,167.9	6,780	15,346.7	1,632.3	93,926.9
	Total	199,595	6,260	24,956.5	509.71	231,321	6,780	45,699.2	89,229.4	373,029

Source: Central Electricity Authority Ministry of Power, Government of India.

TABLE A3.2 Power projects currently operational in the NER (as of 31 March 2019)

Sr. no.	State	Name	Sector	System	Type	Installed Capacity (MW)	Net Generation (GWh)
1.	Arunachal Pradesh	Nuranang	State	Arunachal	Hydro	6	–
		Pare	Center	NEEPCO	Hydro	110	345
		Ranganadi	Center	NEEPCO	Hydro	405	1,047
		Tago	State	Arunachal	Hydro	5	–
2.	Assam	Bongaigaon	State	ASEB	Thermal (Coal and Oil)	0	0
		Bongaigaon TPP	Center	NTPC	Thermal (Coal and Oil)	750	2,574
		Chandrapur_ Oil	State	ASEB	Thermal (Gas)	0	–
		DLF	Private	EIPL	Thermal (Gas)	15.5	–
		Kathalguri GT	Center	NEEPCO	Thermal (Gas)	291	1,596
		Lakwa GT	State	APGCL	Thermal (Gas and Diesel)	212	841
		Mobile Gas T-G	State	ASEB	Thermal (Gas)	0	–
		Namrup GT	State	ASEB	Thermal (Gas)	0	–
		Karbi Langpi	State	ASEB	Hydro	100	371
		Kopili	Center	NEEPCO	Hydro	200	1,112
3.	Manipur	Leimakhong DG	State	MPDC	Thermal (Diesel)	36	–
4.	Meghalaya	Loktak	Center	NHPC	Hydro	105	600
		Khandong	Center	NEEPCO	Hydro	50	203
		Kyredemkulai	State	MEGEB	Hydro	60	134
		Myntdu	State	MEECL	Hydro	126	361
		New Umtru	State	MEECL	Hydro	40	179
		Umiam I, II & IV	State	MEGEB	Hydro	96	250
		UMTRU (NEW)	State	MEGEB	Hydro	11	–

	State	Project	Ownership	Developer	Type		
5.	Mizoram	Tuirial	Center	NEEPCO	Hydro	60	–
6.	Nagaland	Doyang	Center	NEEPCO	Hydro	75	230
		Likim RO	State	Nagaland	Hydro	24	–
7.	Sikkim	Chuzachen	Private	Gata Infra Ltd.	Hydro	110	415
		Dikchu	Private	SKPL	Hydro	96	460
		Jorethang Loop	Private	Dans Energy P Ltd.	Hydro	96	408
		Lower Lagyap	State	Sikkim	Hydro	12	–
		Moyagchu	State	Sikkim	Hydro	4	–
		Rangit-III	Center	NHPC	Hydro	60	–
		Tashiding	Private	Dans energy P ltd	Hydro	97	347
		Teesta-III	State	Teesta Urja Ltd.	Hydro	1,200	4,237
		Teesta-V	Center	NHPC	Hydro	510	2,688
		U. Rognichu	State	Sikkim	Hydro	8	–
8.	Tripura	Agartala GT	Center	NEEPCO	Thermal (Gas)	135	626
		Baramura	State	TSECL	Thermal (Gas)	42	174
		Monarchak CCPP	Center	NEEPCO	Thermal (Gas)	101	656
		Palatana CCPP	Private	Th. Powertech Corp. Ltd.	Thermal (Gas)	726	4,516
		Rokhia GT	State	TSECL	Thermal (Gas)	63	438

Source: CEA, (2020)

TABLE A3.3 Major functional Land Custom Stations (LCSs) in North East India and neighboring countries

Sl.No.	India		Neighboring Country	
	State	LCS	LCS	Country
1	Assam	Gauhati Steamerghat		Bangladesh
2		Dhubri Streamerghat	Rowmati	Bangladesh
3		Mankachar		Bangladesh
4		Silghat		Bangladesh
5		Karimganj Steamer Ghat	Zakiganj	Bangladesh
6		Sutarkhandi	Sheola	Bangladesh
7		Hatisar		Bhutan
8	Meghalaya	Mahendraganj	Dhanua Kamalpur	Bangladesh
9		Dalu	Nakugaon	Bangladesh
10		Baghmara	Bijoypur	Bangladesh
11		Borsora	Borosora	Bangladesh
12		Shellabazar	Sonamganj	Bangladesh
13		Dawki	Tamahil	Bangladesh
14	Tripura	Agartala	Akhaura	Bangladesh
15		Srimantapur	Bibir Bauar	Bangladesh
16		Khowaighat	Balla	Bangladesh
17		Manu	Chatlapur	Bangladesh
18	Mizoram	Demagiri	Rangamati	Bangladesh
19		Zokhawthar	Rih	Myanmar
20	Manipur	Moreh	Tarnu	Myanmar
21	Sikkim	Sherathang (Nathu La)	Renginggang	China
22	West Bengal	Petrapole (ICP)	Benapole	Bangladesh

Source: Authors' own

would help facilitate industries in the BEZs. At the same time, countries need to attract greater investments in power generation and transmission infrastructure.

A look at the GMS experience on the cross-border electricity trading mechanism could further strengthen the case for replicating it for NER–Myanmar in the context of BEZ development, which has immense untapped potential. GMS countries of Cambodia Lao PDR, Myanmar, Thailand, Viet Nam, and China are characterized by uneven load demand and different energy resource base. An Electric Power Forum (EPF) was constituted in 1995 under the GMS economic cooperation program to serve as advisory body on subregional power projects and issues. With the objective of

TABLE 3.10 Renewable energy capacity in the NER to be achieved by 2022 (MW)

States	Target to be achieved by 2022 (MW)	Current installed capacity as on 30.04.2020 (MW)	Balance to be achieved in the next 2 years (MW)
Arunachal Pradesh	239	136.72	102.3
Assam	697	75.34	621.66
Manipur	110	11.05	98.95
Meghalaya	192	46.45	145.55
Mizoram	92	37.99	54.01
Nagaland	122	31.67	90.33
Tripura	239	25.42	213.58
Sikkim	88	52.18	35.82
NER	1779	416.82	1362.2

Source: CEA (2018)

promoting efficient regional power trade, the following guiding principles are recognized by each GMS country.

- Each member recognizes and endorses international trading in energy to be an integral part of its policies to strengthen its electricity sector.
- Each member recognizes the importance of technical harmonization of electric power transmission parameters and practices with eventual interconnection in mind.
- Each member recognizes the desirability of foreign direct investment on reasonable terms in its electricity power sector in order to speed economic development.

7 Key Recommendations and Way Forward

BEZ development and energy market integration in the NER with neighboring economies, particularly Myanmar, a gateway member of the ASEAN, has immense potential, but is in an evolutionary stage, awaiting structured approaches. For the NER to become a significant power hub through establishment of BEZs, the following major recommendations are suggested.

7.1 Energy Integration

7.1.1 Cross-Border Energy Linkages

India has been supplying a small amount of electricity (3 MW) to Myanmar across the Moreh–Tamu border since April 2016. Unlike Moreh–Tamu, Agartala and Chittagong or Shillong and Sylhet do not have a formal energy exchange nor are their grids interconnected, and in some cases grids do not have the needed capacity for energy exchange on a sustainable basis. However, if the regional electricity market comes into effect, cross-border electricity exchange will be smooth and would help BEZs to grow fast. At the same time, it will generate investment opportunities for the private sector in power generation and transmission infrastructure in the region. Successful bilateral energy trading agreements that are in place between Bhutan and India, India and Nepal, and India and Bangladesh offer insights for developing market-based approaches and necessary commercial frameworks

7.1.2 Regulatory Convergence and Harmonization

Convergence in regulatory matters of the energy sector of the participating countries is needed to optimally gain from the energy integration. Countries can gain from electricity trading, particularly to promote the BEZs. However, market integration must be accompanied by domestic reforms and international harmonization of regulatory standards.[19] A number of initiatives from regulatory to legal frameworks have to be designed to encompass technical aspects, such as rules and procedures concerning transmission access and its pricing, congestion management, operational codes and protocols for system operation, energy accounting, wheeling fees, etc. Data and financial transfer protocols need to be gradually harmonized through appropriate regulations for the seamless stable integration of power markets for enhanced energy trade.

7.2 Connectivity Development

Deeper power market integration in the NER and ASEAN with BEZ will be driven by the economy of scale that would in turn depend on other infrastructure developments, as well as multilateral agreements on the free flow of goods, services, and human resources.

7.2.1 Improvement of Road Infrastructure, Completion of TH, and Replacement of 69 Bridges

The road between Imphal and Moreh should be made into a six-lane road. In particular, the Moreh–Pallel section of the road has to be improved. The road in the Monwya–Yargi section in Myanmar should be widened to four lanes. Road conditions in Manipur, particularly those connecting neighboring countries, should be made high quality. Timely completion of the Trilateral Highway (TH) and replacement of 69 bridges is critical to the NER's linkages with Southeast Asia and vice versa. At present, the 112 km road of TH is under construction under the supervision of the National Highway Authority of India (NHAI). The project commenced on 28 May

2018 and is expected to be completed by 2025. Without the completion of the bridges, the TH cannot be made operational for cargo vehicles and passenger bus services between India and Myanmar.

7.2.2 Completion of Negotiation of Trilateral MVA

Progress in the negotiation of the Motor Vehicle Agreement (MVA) between India, Myanmar and Thailand for the TH is slow. Given that all the three countries have ratified the WTO Trade Facilitation Agreement (TFA), TH-countries may resume the MVA negotiation at the earliest and complete the negotiation before the TH comes into operation. In many areas, the WTO TFA and TH MVA are interrelated. Myanmar's progress in implementing WTO TFA has been slow. Myanmar needs technical assistance and capacity building while implementing the WTO FTA. Both India and Thailand may offer adequate technical assistance and capacity building to Myanmar while implementing the TH MVA.

7.3 Financing

Mobilizing cost-efficient investment in energy supply will be a constant challenge for policy makers at national and state levels. Several studies estimate India requires a cumulative US$ 2.8 trillion in investment toward sustainable energy transition. It requires effective coordination between multiple institutions, private sector, and levels of government.

There are two options: first, we follow the Public–Private Partnership (PPP) model, where BEZs (or industrial parks or SEZs) are developed and run by the private sector but the government continues to play a key role in providing legal and infrastructure services. Private investments are welcome for the establishment of supporting services such as bonded warehouse, logistics and distribution centers, service-related activities such as hotels, banks, hospitals, etc. Driven by the private sector, most of the BEZs in the Mekong subregion have a strong tourism component. The NER is one of India's most important tourist destinations. Developing BEZs would facilitate tourism automatically. Engaging the private sector in BEZs thus offers substantial merit. The big challenge is to secure financing for development of BEZs. Public funds may not be adequate to meet this huge investment, so PPPs should be encouraged. An important role for cross-border funding exists, including by multilateral banks and possible new institutions. The State Bank of India and EXIM Bank of India could be important sources for funding the development of BEZs or their components. A reduction in risks facing international investors on new large-scale renewable energy generation and transmission projects will make those projects competitive and enable further expansion of cross-border energy trade.

7.4 Supply Chain Management

7.4.1 Strengthening Backward Linkages

Many BEZs are unable to cope with the rising demand for finished products due to slow or negligible backward linkages. Backward linkages would also depend on the specialization of border industries. Countries have to improve domestic capacity so that local firms from the NER can take part in regional supply chains. Skilling and education are essential for value chain upgradation.

7.4.2 Trade Facilitation and One Stop Service (OSS)

In future, when the RCEP is implemented, communications and connectivity for raw materials, finished goods and services, the supply chain will increase.[20] In such a scenario, investment potential in BEZs along the India–Myanmar border is very high. This One Stop Service (OSS) shall be adopted for investors, and government should provide supporting services for the development of infrastructure of BEZs.

7.4.3 Safe and Secure Borders

Smart borders are essential for security and the safety of goods, vehicles, and passengers. In order to have complete vigilance, border posts have to be equipped with modern gazettes such as scanners, container handling equipment, 24 × 7 security, biometric measures, etc. Simple border-crossing procedures with online transaction are essential to encourage cross-border trade and investment.

7.5 Promotion[21]

- Successful BEZs in the Mekong subregion bring in all stakeholders – the private sector, non-governmental organizations, developers, government agencies – at all stages of development. To facilitate the development of BEZs, the government may consider setting up the Border Economic Zone Development Authority of India (BEZDAI), which will facilitate BEZs in NER. BEZDAI will do the coordination and be responsible for the development of BEZs.
- Initiate a dedicated fund in the form of a special purpose vehicle exclusively to finance power projects in the NER with an objective to facilitate the BEZs.
- Sensitize and build capacity amongst the main development actors, including the political leadership, bureaucracy, and technocrats. In addition to this, organize several seminars and webinars, conferences, and

workshops for the stakeholders, including government institutions, the private sector, civil society members, media, academics, and grass-roots community leaders.

- Bring together central and state agencies in the power sector into a common forum like the Ministry for Development of North Eastern Region and North Eastern Council, and sensitize and train them in the wider dynamics and potential of cross-border power exchanges.
- The North East Industrial Development Schemes 2018 should be revised further by enhancing the coverage of incentives for generating plants with capacities up to 50–100 MW, and allowing foreign investors to participate under its provisions. It is also vital to coordinate with foreign policy-making institutions like the Ministry of External Relations and Ministry of Commerce so that the complex dynamics of cross-border electricity exchanges with neighboring countries and subregions can be handled smoothly and effectively.

7.6 Coordination

- Coordinate closely with a cross-section of agencies that build infrastructure, including roads, railways, power transmission lines, digital communications, and waterways. For instance, the carrying capacities of roads and bridges in the highlands must be planned in such a way that all of this physical infrastructure can both withstand and sustain the movement of heavy equipment and machinery required for generation plants, transmission set-ups, and distribution networks, while absorbing shocks like natural disasters.
- Development of BEZs and energy trading across borders requires close coordination with central and state agencies, multilateral donors and development partners, foreign countries, financial institutions, etc.

7.7 Next Steps

- To move ahead with BEZ development, the Government of India may conduct a detailed feasibility study with the help of ERIA on the energy integration plan. This study shall look into the technical feasibility with a detailed investigation of BEZs as well as energy linkages. Stakeholder consultations, particularly with industry associations, would be needed for effective planning of the zones.
- The Ministry of Northeast Development (DONER), Government of India may consider setting up a Joint Working Group (JWG) and a Task Force for coordination with the state governments, and for the development of BEZs.
- The Ministry of External Affairs (MEA), Government of India may set up a team for inter-country coordination.

8 Concluding Remarks

Border barriers have been declining in India's North Eastern Region in the present era of Look East–Act East Policy. Three BEZs have been suggested in this study for development in North East India: (i) Manipur (Moreh–Tamu border) with the Manipur and Myanmar border (Moreh–Tamu area), Meghalaya–Bangladesh (Shillong–Sylhet area), and Tripura–Bangladesh (Agartala–Chittagong area). This study argues that the development of border areas through BEZs will help us realize a balanced development and bring the NER from periphery to the core of today's development process. The BEZs of the Mekong subregion offer many important lessons to enhance border connectivity in India, and also between India and Southeast Asia and Bangladesh as well as energy sector integration opportunities. Myanmar has received the GSP benefits, like Cambodia or Lao PDR in Mekong. India may also gain huge benefits if BEZs are promoted, particularly with Myanmar.

A repositioning of the NER to utilize its rich and unharnessed renewable resources effectively is required. BEZs will also provide opportunities to MSMEs. While cross-border value chains may also pick up the pace through BEZs when quality energy supply is ensured, participating countries may consider comprehensive incentives for investments in the energy sector.

Countries participating in BEZ development have to work together to harmonize regulations and technical standards in the energy sector. At the same time, the Indian Power Ministry, for example, may consider identifying regional best practices and meeting with global peers in the energy sector.

A number of successful international experiences clearly suggest that investments in cross-border energy trade, complemented by policy and regulatory arrangements can help not only address the energy security but also provide a positive thrust to the local economy. International experiences further strengthen the argument, and in the long run support a regional framework to promote investments in developing cross-border clean energy projects and developing a competitive power market.

Globally, many regional and other international energy trade/cooperation methods have been sustained through regional regulatory forums such as the Agency for Energy Cooperation (ACER), Council of European energy regulators (CEER), Energy Regulators Regional Association (ERRA), Head of ASEAN utilities Association (HAUPA), etc. These regional institutions have been helpful in coordinating or harmonizing national policy frameworks to promote cross-border energy trade and related increments.

Finally, these key recommendations should be studied further to generate more information and knowledge, create critical space for cross-border energy exchanges, generate adequate alternatives for policy interventions, and highlight the scope for cooperation and integration with Southeast Asian countries. A deeper and wider assessment of all of the crucial findings in this study could attract significant interest from national, regional, global, and multilateral investors for the development of BEZs in NER.

Notes

1 These include the South Asian Free Trade Agreement (SAFTA), the Bay of Bengal Initiative for Multi-Sectoral Technical and Economic Cooperation (BIMSTEC) and the Asia-Pacific Trade Agreement (APTA).
2 This quadrangular perspective covers (i) within a state, (ii) amongst the NER states, (iii) with the rest of India, and (iv) cross-border interactions, and will bring openness, reoriented thinking, and varied opportunities for the NER (refer, Anbumozhi et al., 2019).
3 Economic corridors and BEZ are inter-linked.
4 Based on the definition outlined in the Ministry of Commerce and Industry, Government of India.
5 Refer to the joint statement issued during the visit of Indian Prime Minister to Bangladesh on 27 March 2021, www.mea.gov.in/bilateral-documents.htm?dtl /33746/.
6 Refer, for example, to Kudo and Ishida (2013), Ishida (2009).
7 Ibid.
8 Refer, for example, to Mekong Institute (2018).
9 In literature, BEZs are also described as SEZs, industrial parks, bonded zones, border trade zones, etc.
10 An impressive US$ 15 billion or so invested in GMS projects (ADB, 2017).
11 Based on several secondary information, primarily drawn upon ADB (2016), Mekong Institute (2018), Kudo and Ishida (2013), AIC (2017), etc.
12 Noted in Tsuneishi (2007), Mekong Institute (2018).
13 During 2011–2012 and 2015–2016, ass total exchange of Rs. 168.66 million was carried out in these four Border Haats (sourced from the Ministry of Commerce and Industry, India).
14 Based on the field visit to industrial estates maintained by the Manipur Government by author during 7–11 May 2019.
15 Refer, for example, Sarma and Choudhury (2018), De and Majumdar (2014), Chong (2018), De and Kunaka (2019), Sarma and Bezbaruah (2009)
16 Banerjee (2020) has called for converting Border Haats into cross-border international retail trade zones.
17 A series of such measures has been laid down at ERIA (2019).
18 Refer, NEPCO, available at https://neepco.co.in/projects/power-potential Contribution of NEEPCO in the hydro installed capacity of NER is 1,225 MW i.e. about 70.93%.
19 This was also raised in Wu et al. (2012).
20 RCEP excluding India.
21 Based on Anbumozhi et al. (2019).

References

Anbumozhi, V., Kutani, I., & Lama, M. P. (2019). *Energising connectivity between Northeast India and its neighbours.* Jakarta: ERIA.

ASEAN-India Centre (AIC). (2017). *Mekong-Ganga cooperation: Breaking barriers and scaling new heights.* New Delhi: Research and Information System for Developing Countries (RIS).

Asian Development Bank. (2016). *Role of special economic zones in improving effectiveness of GMS economic corridors*. Manila: Asia Development Bank (ADB).

Banerjee, P. (2020). Reimagining border Haats as international retail trade zones, Discussion paper, CUTS, Jaipur.

Brunner, H. P. (2010). *North East India: Local economic development and global markets*. New Delhi: SAGE Publications.

Chong, B. (2018). Pursuing development through connectivity: An analysis of India's northeast region, Centre on Asia and, Globalisation, Lee Kuan Yew School of Public Policy, National University of Singapore, Discussion paper, Retrieved from July 24, 2019.

De, P., Dash, P., & Kumarasamy, D. (2020). The trilateral highway and Northeast India: Economic linkages, challenges, and the way forward. In *The India–Myanmar–Thailand trilateral highway and its possible eastward extension to Lao PDR*. Cambodia, and Viet Nam: Challenges and Opportunities-Background Papers. ERIA Research Project Report FY2020 no.02b, Jakarta: ERIA, pp. B1-1–46.

De, P., & Kunaka, C. (2019). Connectivity assessment: Challenges and opportunities. In S. Kathuria & P. Mathur (Eds.), *Playing to strengths: A policy framework for mainstreaming Northeast India*. Washington, DC: The World Bank.

De, P., & Majumdar, M. (2014). *Developing cross-border production networks between north eastern region of India, Bangladesh and myanmar: A preliminary assessment*. New Delhi: Research and Information System for Developing Countries (RIS).

EXIM Bank. (2018). Act east: Enhancing India's trade with Bangladesh and myanmar across border, Working Paper No.77, Export-Import Bank of India, Mumbai, Retrieved from August 3, 2019.

Ishida, M. (2009). Special economic zones and economic corridors. In A. Kuchiki & S. Uchikawa (Eds.), *Research on development strategies for CLMV countries*. ERIA Research Project Report 2008-5 (pp. 33–52). Jakarta: ERIA.

Ishida, M. (2013). Epilogue: Potentiality of border economic zones and future prospects. In M. Ishida (Ed.), *Border economies in the greater Mekong sub-region*. Basingstoke: Palgrave Macmillan, pp. 299–331.

Kudo, T., & Ishida, M. (2013). Prologue: Progress in cross-border movement and the development of border economic zones. In M. Ishida (Ed.), *Border economies in the greater Mekong sub-region*. Basingstoke: Palgrave Macmillan, pp. 3–28.

Mekong Institute. (2018). Joint Study and Survey of Special Economic Zones (SEZs) and Cross Border Economic Zones (CBEZs) to match complementary SEZs and identify prioritized areas. Khon Kaen.

Murayama, M., Hazarika, S. & Gill, P. (2022) Northeast India and Japan: Engagement through Connectivity. New Delhi: Routledge.

NITI Aayog. (2018). NITI forum for North East. New Delhi. Retrieved from https://niti.gov.in/niti-forum-north-east.

Sarma, A., & Bezbaruah, M. P. (2009). Industry in the development perspective of Northeast India. *Dialogue, 10*(3), 55–64.

Sarma, A., & Choudhury, S. (2018). *Mainstreaming the northeast in India's look and act east policy*. Singapore: Palgrave Macmillan.

Singh, P. (2020). *The missing link: Northeast India in the context of TH and development*, Report Prepared for the ERIA Study on Trilateral Highway, RIS, New Delhi.

Tsuneishi, T. (2007, August). Thailand's economic cooperation with neighbouring countries and its effects on economic development within Thailand, IDE Discussion Paper No. 115 (pp. 1–29).

UNESCAP (2018) Integrating South Asia's Power Grid for a Sustainable and Low Carbon Future, Bangkok.

USAID (2016) South Asia Regional Initiative for Energy Cooperation and Development. Washington, D.C.

Wang, Z. (2016). Cross-border economic zone as strategy for economic corridor development: Concept, rationale and driving forces in the border areas of PR China, Presentation made on October 21, 2016 at Yunnan University, Kunming.

Wu, Y., Shi, X., & Kimura, F. (2012). Energy market integration in East Asia: Theories, electricity sector and subsidies, ERIA Research Project report 2011, No. 17, ERIA, Jakarta.

Yoshikawa, H., & Anbumozhi, V. (2017). *Electricity futures in the greater Mekong subregion: Towards sustainability, inclusive development, and conflict Resoution.* Jakarta: Economic Research Institute for ASEAN and East Asia.

Appendix 3

Based on the consultation that the author had with the Manipur Government and other stakeholders during the field visit on 7–11 May 2019.

Manipur Chamber of Commerce and Industry

- Trade and violence cannot go together. Seamless movement of goods between Moreh and Dimapur and Moreh and Silchar is very much needed.
- Law and order is the issue for peace and prosperity in Manipur.
- Problems faced by local traders and problems faced by manufacturers are different.
- We have a very good industrial policy North East Industrial Development Scheme (NEIDS). But, we need incentives for construction of buildings and not just plants and machinery.
- We need to prepare connectivity and people-to-people contacts.
- We need to study not only the Moreh–Tamu border but also other borders in Myanmar. We should analyze the border management carefully.
- Police harassment on the Indian side of the border is very high.
- Improvement of infrastructure and the basic amenities in Tamu and Moreh is needed.
- We need to draw lessons from borders elsewhere.
- Presently, it is all about cash transactions between Indian and Myanmar traders.
- We need to move from informal payment to more formal payment methods through bank transfers.

- Myanmar should accept third-country goods, particularly Thai products.
- Myanmar should also improve basic amenities at the Tamu side of the border.

Government of Manipur

- The trade should happen through (letter of credit) banks and not through informal channels.
- The Manipur Government is keen to develop Border Haats with Myanmar. Face-to-face interactions between the authorities for development of border huts is needed.
- The Government of India should reduce the tariff hike.
- On 25 April 2019 there was a letter from the Prime Minister's Office (PMO) requesting the enhancement of the India–Myanmar border trade. However, discussion is needed on the modalities of trade between the two countries.
- The Manipur Government has taken steps to set up BEZs in the Manipur state such as food parks, industrial estates, and townships, etc.
- The Manipur Government is planning to set up empowered teams for facilitation of trade, people-to-people contacts, and economic interactions between the Manipur and Myanmar Government under the overall guidance of the Government of India through its Act East Policy.

4

ENERGY MARKET AND HIERARCHICAL INTERACTIONS BETWEEN INDIA'S NER AND THE ASEAN – FROM THEORY TO PRACTICE

Potential of Energy Trade in Conventional and New Energy Resources; Market Preferences and Regulatory Choices; Policy Implications for the NER and the ASEAN

Kakali Mukhopadhyay, Priyam Sengupta, and Vishnu S. Prabhu

1 Introduction

Since the 1990s, India has heavily pursued building connectivity with its eastern neighbors to access the vibrant markets of the Association of Southeast Asian Nations (ASEAN). Using the land borders of the North Eastern Region (NER) of India as a gateway to ASEAN countries was always seen as a strategic move on the part of the Government of India, which was also the premise in its landmark external policies, such as the "Look East Policy" (now "Act East Policy"). Quite interestingly, the role of cross-border power trade turned out to be very crucial for economic integration between India and its eastern neighborhoods, including the ASEAN. More specifically, unleashing the renewable energy potential of the NER can emerge as a game changer in this process.

The objective of this chapter is to assess the importance of hierarchical interactions across all stakeholders for cross-border energy trade, the role of hydropower as a grid stabilizer and importance of cross-border grid integration with the regional power grid (ASEAN Power Grid) as a policy instrument. Some allied issues like the Government of India's latest policy of infusing more renewable energy sources (RES) in the electricity system of the country for achieving the dual objective of reducing air pollution and cost optimization have also been discussed at length.

DOI: 10.4324/9781003433163-4

The concept of hierarchical interactions in decision-making and the energy hierarchy has been comprehensively explained in Section 2. The NER's comparative advantage in hydropower generation as well as other energy sources are discussed in Sections 3 and 4, respectively. Further, cross-border power system integration linking the NER and ASEAN countries is also touched on in Section 4. Section 5 discusses the potential energy trade between the NER and ASEAN. Section 6 delineates the prospect of cross-border trade between India's NER and its neighboring countries, such as ASEAN countries, and the chapter ends with a conclusion and policy recommendations in Section 7.

2 Importance of Energy Hierarchy and Interactions with All Relevant Stakeholders for Cross-Border Trade

Due to the dependence of Variable Renewable Energy (VRE) on the existing weather conditions during a particular point in time, its integration in the electricity grid remains vulnerable. In such cases, the usage of Smart Energy Systems which take into consideration various intra-hour, daily, seasonal and biannual storage options create higher grid flexibility, and thus accommodate a higher penetration of RE sources (Mathiesen et al., 2015). Complementing Smart Energy Systems with the existing smart grid research on smart metering and potential power storage technologies will not only increase the adoption of VRE, but also lead to higher energy efficiency in the power sector. Furthermore, in a smart grid system, consumers can become prosumers (producers and consumers) and trade surplus energy generated from, for example rooftop photovoltaic system over the power grid (Ilic et. al., 2012). Thus, market-driven hierarchical interactions can form smart grid neighborhoods for electricity trade. With the evolution of energy systems, the decision-making hierarchy consisting of environmental, economic, social and governmental dimensions should be inclusive of all relevant stakeholders (Stranzali & Aravossis, 2016). Tziogas et al. (2019) prepare a hierarchical analysis framework to fasten the transition toward sustainable electricity systems. The authors suggest that various economic, social and environmental sustainability indicators can assist in the design and management of electricity systems using a System Dynamics tool to capture the impact of the non-linear nature of electricity demand. With efficient electricity forecasting, appropriate policy interventions can be made to reduce the burden on operational performance of RE sources. One significant aspect of adopting a System Dynamics approach is the feedback mechanism it provides which captures the real patterns of energy systems' behavior to non-linear changes in electricity demand over time, thus assisting in improving effectiveness through efficient decision-making (Roberts, 1978).

The energy hierarchy is determined by the prioritizing of energy options available for electricity generation, consumption and trade depending upon the technical, socio-economic, environmental and technological factors. La Rovere et al. (2010) conducted a Multicriteria Analysis (MCA) based on the above-mentioned factors and showed that a small hydropower project can be put at the top of the energy hierarchy thus ensuring grid flexibility and reliability, compared to wind energy, sugarcane bagasse, biodiesel, urban solid wastes, natural gas and nuclear energy. However, the ranking of energy sources may vary depending upon the changes in the degree of factors affecting the particular project portfolio. Ahmad and Tahar (2014) conduct an analytical hierarchy process where they find that hydropower has a comparative advantage in the technical aspect, whereas solar for economic, biogas for social and wind power for environmental aspect.

Verbong and Geels (2010) find that electricity over the past many decades has been dominated by an energy hierarchy in which the priorities in decreasing order are cost efficiency, system reliability and lastly environmental issues. The primary goal of the electricity generators is to ensure the demand is met at cost-effective rates, in cooperation with the state distribution companies and maintaining grid reliability for which fossil fuel sources are largely utilized. Only once there is strong physical infrastructure to meet the first two priorities can the adoption of RE sources for electricity generation be taken into consideration. However, today, with higher emphasis for adoption of cleaner energy sources in the electricity mix, the authors classify the energy hierarchy, in decreasing order of priorities, as system reliability, environmental issues and, lastly, cost effectiveness. Such an order is necessary to ensure the energy security of the nation by implementing reformative regulatory interventions, necessary to incentivize producers in market regimes..

In Figure 4.1, we can see the various social, economic, institutional and environmental factors involved in the decision-making process across various government and company-level stakeholders.

3 Importance of Hydropower Generation in the NER from the Perspective of Grid Stability

The Government of India's Ministry of New and Renewable Energy recently announced the target of 70 GW hydropower capacity to be achieved by 2030 (Energetica India, 2020). This decision acquired strength after the revelations of the "lights-out" event on 5 April 2020, when the importance of hydropower came into focus in a big way. Following Prime Minister Narendra Modi's appeal, Indian citizens all across the country displayed their collective strength and resolve to fight against COVID-19 by turning off lights for 9 minutes at 9 pm on 5 April 2020 (Kumar & Krishnan, 2020). The

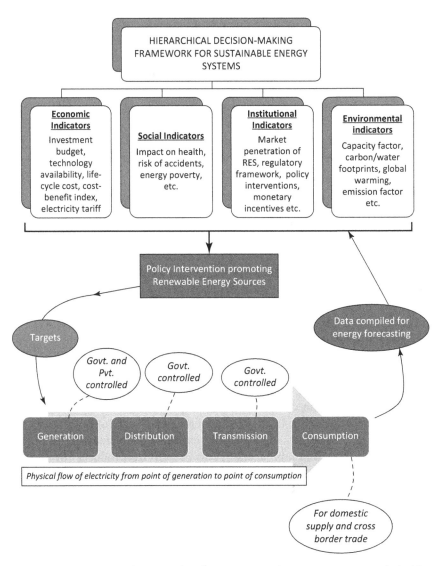

FIGURE 4.1 Conceptual Hierarchical Interactions between various stakeholders in the energy sector.

Source: Created by Authors

country's power companies were highly apprehensive because sharp fluctuations in supply had potential to destabilize the grid. But during the lights-out event, fluctuation of power demand was managed with minimal issues and with judicious use of small hydro plants. The important fact that technically hydropower provides maximum flexibility to the grid operator, was proved beyond a doubt during this event (Joshi, 2020). The power ministry in its draft amendment of the Electricity Act 2020 has already proposed the inclusion of hydropower within the Renewable Purchase Obligation (RPO) undertaken between state power distribution companies and power generators (NEEPCO, 2017).

To achieve the national target of 70 GW installed capacity of hydropower by 2030, the North Eastern states of India need to play a pivotal role. The NER, composed of Sikkim, Arunachal Pradesh, Assam, Meghalaya, Manipur, Mizoram, Nagaland and Tripura, is endowed with 40% of the total hydropower potential of the country (NEEPCO, 2017). The NER is blessed with a huge hydro potential of about 58.9 GW, out of which 1.9 GW has been harnessed as of November 2021 which is only 3.2% of total hydro potential. An additional 2.3 GW of hydropower projects are under construction. Nevertheless, hydropower continues to be the largest source of energy supply in the region, constituting 39.7% of total installed capacity, followed by 36.3% natural gas and only 15.7% by coal-fired thermal power plants (TPPs) (CEA, 2021). In contrast, the rest of India's energy system is largely based on the use of coal-fired power plants for electricity generation, which is largely domestically supplied, and oil and natural gas for transport and industry which is highly imported. But usage of fossil fuels in producing power has its usual drawbacks. First, it causes too much environmental pollution. Second, the coal stock is continuously depleted, which may ultimately cause the need for import in future. To avoid such a situation, non-renewable sources of energy must be supplemented well by renewable forms of energy.

4 Hydropower as a Variable Renewable Energy Source in the Energy Hierarchy

The grid stability in India is maintained by the Central Electricity Regulatory Authority (CERC), which manages any deviation from pre-determined frequency by varying power flow in the grid. Sudden fluctuations in frequency can cause a collapse. As the share of electricity produced from VRE sources grows, so does the need for its integration in a cost-effective manner. Increase in power generation from VRE will also ensure higher grid flexibility resulting from the upgradation of energy hierarchy in the region since there will be a number of alternatives through which electricity demand can be met. Furthermore, reliance on VRE will also lead to positive environmental gains as well, resulting from low CO_2 emissions from electricity generation vis-à-vis fossil fuel sources. Electricity generation from hydropower is one of the cleanest sources of energy offsetting high levels of CO_2 emissions if it can substitute coal TPPs on a large scale.

We can see from Table 4.1 that CO_2 emissions from hydro are only 1% of emissions resulting from coal-fired TPPs. However, large hydropower projects have higher capital costs estimated at USD 1.4 million/

TABLE 4.1 Source-wise GHG emissions per unit of electricity produced in India

Sr. no.	Source of Power	GHG emissions (gCO₂e)
1.	Thermal – Coal	957
2.	Thermal – Gas	422
3.	Solar	38
4.	Hydro – Storage	10
5.	Wind	9
6.	Nuclear	6
7.	Hydro – RoR	4

Source: GOI, (2019)
Note: RoR: Run-of-River

MW (INR 10 crores/MW) compared to USD 1.14 million (INR 8 crores/MW) for a coal-fired TPP project as well as longer gestation period (GOI, 2019). Furthermore, an electric system with an increasing share of RES will act as a catalyst in transforming the transmission and distribution (T&D) segment of the power sector as well. Accordingly, power utilities and grid operators must invest in upgrading the T&D sector which requires time due to the sophistication involved in learning the technological know-how to ensure system reliability. In the short run, VRE sources such as solar need to be complemented with fossil-fuel sources as a consequence of variation in output resulting from unforeseen weather conditions (Weiss & Tsuchida, 2015).

Since in the NER hydroelectricity is mostly generated from rain-fed perennial streams flowing through mountainous topography, the hydroelectricity produced here is highly variable in nature. Thus, sudden variation in this VRE can pose serious threat to the grid stability in the region. The standard solution prescribed to avoid such a problem in grid stability (due to grid integration of VRE into the system) is to ensure sufficient energy options to enable the grid operator to switch from one to another option, depending on situations. To accomplish this task of grid management most efficiently, ascertaining energy hierarchy is a pre-requisite.

The act of cross-border power system integration involves a wide range of elements: cooperation on system planning; grid synchronization; co–ordination of system operations; integration of electricity markets; and harmonization (or consolidation) of policies and regulation. On a closer look, it is evident that ascertaining hierarchical interactions is the most vital part of cross-border power trade between India and the ASEAN. During cross-border energy trade, the hierarchical interactions could happen at two levels: at the country level or at the company level, and they can occur along a spectrum of cross-border energy; for example, market transactions on one end compared to coordinated actions by the grid operators of the two countries on the other end. Joint action plans by the governments (e.g., ASEAN

Energy Cooperation Action Plan, Myanmar–India Comprehensive Economic Partnership) and ventures by the grid companies represent an intermediate step, whereby the two parties commit resources to a new project that they co-own and co-manage. This chapter will try to describe the incidence of such interactions linking the NER and ASEAN countries.

5 Potential of Energy Trade in Conventional and New Energy Resources between the ASEAN and North East States

5.1 Importance of the Land Border with Myanmar

Myanmar provides a gateway for India for extending strategic and economic ties with South East Asia due to the shared land and maritime borders between the countries. India and Myanmar are also members of the Bay of Bengal Initiative for Multi-Sectoral Technical and Economic Cooperation (BIMSTEC), as well as members of the Bangladesh–China–India–Myanmar Forum for Regional Cooperation (BCIM) (Taneja et al., 2019).

Cooperation in the energy sector is always considered to be one of the most important routes to integrate economies with close proximity. For energy market integration, the presence of a common land border is required to enable the cross-border connection between the power grids of two partner countries. In this respect, Myanmar is again an essential strategic partner of India, since it is the only ASEAN nation with which India shares 1,643 km of land border. The Border Trade Agreement between India and Myanmar, signed on 21 October 1994 in New Delhi, provided for the following three Land Customs Stations (LCS) on the Indian side of the border:

1. Moreh in Manipur.
2. Zokhawthar (Champai) in Mizoram.
3. Nampong in Arunachal Pradesh.

In 2015, the Director General of Foreign Trade (DGFT) upgraded the border trade at the Moreh–Tamu checkpoint to "normal trade", implying trade between countries will be subject to payment of custom duties applicable on trade with any other country in the world (Das, 2016). Unlike the simple barter trade that was happening mostly in agricultural commodities, normal trade would encourage export of sophisticated capital goods between the two countries, thus increasing the prospects of trade intensity in the region. The governments of India and Myanmar recently implemented the Land Border Crossing Agreement which was marked by the simultaneous opening of the Tamu (Myanmar)–Moreh (India) and Rihkhawdar (Myanmar)–Zowkhawtar (India) international border checkpoints.

The potential for mutually gainful trade between the NER and ASEAN countries through these LCS (Moreh and Zowkhawtar) are limited. Since the NER and Myanmar share similar economic and geographic structures, commodities produced are hardly complementary in nature and lack any solid basis of mutually beneficial trade. The prospect of power trade seems to be comparatively better through land borders due to the close proximity between the NER and Myanmar.

5.2 Current Status of Power Trade between India and Myanmar

There is good potential for power trade between India and Myanmar due to the close proximity between the regions. In 2016, an 11 KV interconnection for supplying 2–3 MW electricity was commissioned between Moreh to Tamu (Embassy of India, Yangon, 2019). Currently, India supplies goods and related services for a 230 KV transmission system which is associated with ThahtayChaung Hydropower Project.

A large part of the population in Myanmar lives in rural areas which are inaccessible for conventional sources of energy. In Myanmar, biomass energy constitutes a major portion of primary energy consumption. In this context, India can extend cooperation with Myanmar in the utilization of renewable energy sources, such as biogas plants and biomass (Ray Chaudhury & Basu, 2015).

5.3 Government-Level Hierarchical Interactions to Foster Energy Cooperation

In 2016, the government delegates from the respective countries agreed to a Joint Working Group at Joint Secretary/Director General level and sub-working groups would be formed in specific areas like Transmission Planning, Renewable Energy, Energy Efficiency, etc.

In 2020, India and Myanmar agreed to cooperate in petroleum refining, stockpiling, blending and retail through a Government-to-Government Memorandum of Understanding (GOI, 2020). Both sides welcomed investments by Indian oil and gas Public Sector Undertakings (PSUs) in Myanmar's upstream sector and agreed to explore opportunities for exporting to India a portion of the output from such projects.

6 Scope of Electricity Trade between the NER and Neighboring Countries

Over the years, India's electricity trade with its adjacent countries has shown an increasing growth rate. Table 4.2 shows that, over the last five years, India has transitioned from being a net importer to a net exporter of electricity: not only has India's import from Bangladesh increased, but India's export has also increased at a faster rate over the last five years.

TABLE 4.2 India's cross-border electricity trade (in MUs)

Sr. no.	Year	Exports				Imports	
		Nepal	Bangladesh	Myanmar	Total	Bhutan	Net exports
1.	2015–2016	1,470	3,654	0	5,124	5,557	–433
2.	2016–2017	2,021	4,420	3	6,444	5,864	580
3.	2017–2018	2,389	4,809	5	7,203	5,611	1,592
4.	2018–2019*	2,478	5,057	6	7,541	4,608	2,933
5.	2019–2020**	1,839	6,168	7.3	8,014	6,165	1,849

Source: Kesavan (2019); ET Energyworld (2020)
Note: *Apr 2018–Feb 2019; **Apr 2019–Jan 2020

TABLE 4.3 Prospective export destinations (in INR/kWh)

Sr. No.	States	Prospective Export Destinations	Average Cost of Supply (2018–19)
1.	Assam	Bangladesh	6.53
2.	Meghalaya	Bangladesh	4.95
3.	Manipur	Myanmar	5.79
4.	Tripura	Bangladesh	4.24
5.	Arunachal Pradesh	Myanmar	7.09
6.	Mizoram	Myanmar	7.46
7.	Nagaland	Myanmar	9.74
8.	Sikkim	Through other states	3.41

Source: PFC (2020) and Authors' Calculation

With different states in the NER generating electricity at different capacities, the scope of cross-border trade will depend upon the combination of proximity vis-a-vis the adjacent countries and the comparative cost advantage of electricity export. Table 4.3 shows which states can be targeted for exploring the prospects of trade amongst the adjacent countries based on geographical proximity and cost advantage.

India's North Eastern Region has to increase electricity generation significantly to meet the huge demand from Bangladesh and Myanmar collectively. This should be in addition to supplying electricity for domestic requirement, which is still suffering from electricity shortage. In the current scenario, due to the abundance of coal deposits not only in the NER but across the country, electricity generation as well as the trade of electricity is cheaper from coal generated thermal plants, compared to any other source.

Table 4.4 shows that the adjacent countries find it cheaper to buy electricity generated by coal-fired power plants compared to any other sources.

TABLE 4.4 Tariff on import of electricity from India

Sr. no.	Options	Tariff (INR/ kWh)	Planned capacity (in MW)	Challenges
1.	LNG-based electricity	6.8–9.5	Over 5,000	Location of regasification units and construction; tariff will fluctuate with LNG prices.
2.	Coal (domestic)	4.1–5.4	2,000–3,000	The government has put a moratorium on open-pit mining.
3.	Coal (imported)	5.4–6.8	More than 10,000	More than 10,000 Country lacks a deep-sea port, challenges in location and investment for plants.
4.	Nuclear	5.4–6.8	2,400 by 2024	Investment and trained manpower.
5.	Solar	7.4–8.1	2,000 +	Land availability.
6.	Power (thermal)	4.1–6.8	3,000 +	Contingent on Indian government policy and surplus capacity in India.
7.	Hydropower	N.A.	3,000 +	Dependent on several factors including final tariff at Bangladesh border after wheeling through India.

Source: Pillai and Prasai (2018). Note: The information was provided in USD cents/kWh. It was converted to INR/kWh with the exchange rate of 1 USD = 68 INR

Thus, the increase in electricity generation can lead to increasing dependence on coal-fired thermal power plants. This provides an economic opportunity at present but has harmful environmental consequences in the current scenario, for the region in future. This challenges the objective of sustainable growth and shift to renewable sources in domestic policy.

7 Inter-Regional Grid Connectivity for Cross-Border Trade

The Indian power grid system is divided into five regional grids. Under the initiative of "One Nation, One Grid" the government integrated all the regional grids into one "National Grid" which has been operational since 2014 (Bhaskar, Raj & Ramanathan, 2014). The synchronization of all regional grids leads to minimal leakages and optimal energy utilization by inter-regional transfer from power surplus to power deficit regions (POWERGRID, 2020). During the 2019–2020 budget session, the Finance Minister Nirmala Sitharaman again reiterated the importance of "One Nation, One Grid" to provide affordable electricity across regions

TABLE 4.5 Inter-regional transmission links and capacity target (2017–2022) (in MW)

Sr. No.	Inter-Regional Corridors	Capacity achieved till FY17	Additional capacity planned between 2017–2022	Cumulative capacity by end of FY22
1.	West–North	15,420	21,300	36,720
2.	North East–North*	3,000	0	3,000
3.	East–North	21,030	1,500	22,530
4.	East–West	12,790	8,400	21,190
5.	East–South	7,830	0	7,830
6.	West–South	12,120	11,800	23,920
7.	East–North East*	2,860	0	2,860
8.	Total	75,050	4,300	118,050

Source: CEA (2019)

Note: *Recently, the government announced "The North Eastern Region Power System Improvement Project". This involves the expansion of transmission lines in the NER by 2,100 km and 2,000 km of distribution links to the power grid which is expected to increase electricity demand in the region by 4,000 MW (Bhaskar, 2020).

(Raghavan, 2019). By the end of 2017, the government achieved an inter-regional transmission capacity of 75,050 MW at the national level. By 2022, it is planned to achieve transmission capacity of 118,050 MW. However, the capacity addition planned between 2017–2022 has not allotted any share to the NER.

In Table 4.5, it can be seen that for the period 2017–2022, there has been no capacity addition in terms of building transmission links to connect the NER with the rest of India which has hindered the scope of increasing connectivity of the NER with its immediate neighborhood. Interconnected grid construction also leads to smoother flow of electricity from the point of origin to the point of destination across regions. The interconnectivity of electricity grids can be extended with the neighboring countries in compliance with the existing cross-border electricity trade regulations. This extension of transmission links can be seen as an extension of India's national grid to adjacent countries. Transmission links can be built with adjacent countries, such that not only electricity can be exported from the NER, but also from other regions via the NER through the interconnected grids.

Under the "The North Eastern Region Power System Improvement Project", new transmission (34) and distribution (50) substations are expected to be constructed in the region for which an estimated USD 957 million has been allocated (Bhaskar, 2020).

Table 4.6 shows that from the existing transmission links through which the NER is connected with the National grid, it is a net importer of electricity

TABLE 4.6 Inter-regional energy exchange (in MUs)

Sr. No.	Year	Net export from the NER to ER	Net export from the NER to NR
1.	2013–2014	–2,110.62	-
2.	2014–2015	–1,909.82	-
3.	2015–2016	–1,134.6	–386.52
4.	2016–2017	–2,097	1,698.34
5.	2017–2018	–3,272.5	3,189

Source: NERPC (2018)

from the Eastern Region (ER) and a net exporter to the Northern Region (NR). This national grid can be extended to Myanmar through which the surplus electricity from other regions of India can be traded via the NER. The future electricity position will provide a guideline for developing trade.

Table 4.7 shows that by 2026–2027, the NER's electricity demand will increase by more than twice what is observed today. Thus, increased electricity generation, mainly from hydropower plants focused on meeting domestic requirements can also enhance the prospects of cross-border electricity trade between the NER and adjacent countries through the extension of the national grid. The availability of an efficient inter-connected electricity grid will help in increasing not only the intensity of electricity trade through the current transmission links but will also open up more opportunities for newer electricity trade projects with the adjacent countries.

Under the "One Sun, One World, One Grid" (OSOWOG) initiative, The Ministry of New and Renewable Energy (MNRE), the Government of India asked for a proposal from RE sources for cross-border electricity (MNRE, 2020). Under this initiative the Prime Minister has called for energy supply through solar power across borders. The initiative broadly covers countries located to the East and West of India, namely Myanmar, Vietnam,

TABLE 4.7 Region-wise increase in peak electricity demand

Sr. no.	Region	2019–2020 (in MW)	2026–2027 (F) (in MW)	% Increase
1.	Northern Region	66,559	91,782	37.90%
2.	Western Region	59,416	94,825	59.60%
3.	Southern Region	53,579	83,652	56.13%
4.	Eastern Region	23,421	35,674	52.32%
5.	North Eastern Region	2,989	6,710	124.49%
6.	All India	205,964	318,043	54.42%

Source: Kumar (2018); CEA (2019a); CEA (2020a)

Thailand, Lao, Cambodia, etc., in the East and the Middle East and Africa in the West, with an eventual goal of creating an interconnected grid across borders for electricity trade. MNRE aims to analyze the demand–supply position of our neighboring countries in the East and West zones, and how those gaps can be filled by building renewable energy capacities by 2030. The long-term goal of MNRE will be to analyze the power market dynamics, identify policy interventions for cross-border (and inter-regional) power trade and build an OSOWOG vision for 2050. Furthermore, the BIMSTEC initiative will help in cooperation of expansion of trade at multi-regional level and through support of international financing.

8 Cross-Border Trade in Hydrocarbon Products

In order to explore hydrocarbon reserve in the NER and to promote cross-border trade through the NER, the Government of India (2015) came up with Hydrocarbon Vision 2030 (DGH & CRISIL, Hydrocarbon Vision 2030 for Northeast India, 2015). The potential of the NER as an exporter of Petroleum Oil and Lubricants (POL) products to the adjacent countries is shown in Table 4.8.

The surplus available can be used to meet the demand of neighboring countries' energy needs. India has been supplying Euro-III diesel in the range

TABLE 4.8 Demand potential in neighbouring countries, million metric tonnes per annum (MMTPA)

Sr. no.	Particulars	2014–2015	2019–2020 (F)	2024–2025 (F)	2029–2030 (F)
Demand potential in neighboring countries					
1.	Myanmar	2.00	1.25	1.00	0.75
2.	Bangladesh	4.00	4.00	3.60	5.40
3.	Nepal	1.40	1.70	2.07	2.52
4.	Bhutan	0.10	0.12	0.15	0.18
5.	Sri Lanka	2.00	1.50	2.50	3.50
6.	Total demand potential for POL products	9.50	8.57	9.32	12.35
Supply potential in the NER					
7.	Production capacity (POL)	6.5	7.0	14.3	14.3
8.	Demand from the NER	3.20	4.0	5.4	7.3
9.	Surplus available	3.3	3.0	8.9	7.0

Source: DGH and CRISIL (2015)

of 10,000–12,000 liters per month to Bangladesh by rail (Bose, 2018). The oil is pumped in Assam by Numaligarh Refinery Limited (NPL) and transported through a pipeline from Siliguri, West Bengal. In 2018, the governments of the two regions also signed an agreement, termed "the friendship pipeline project" in which India will supply diesel through pipeline instead of rail, thus making it more cost effective. The 130 km pipeline has a capacity of 1 Million Metric Tons Per Annum (MMTPA). Bangladesh also finds import of oil by pipeline from India far more cost effective than importing oil from other countries to the Chittagong Sea port and then transporting it to the point of consumption by road. Bangladesh also only recently started its commercial exports of Liquefied Petroleum Gas (LPG) to the state of Tripura with a shipment of 1,000 tons per month (Rahman, 2020).

The mineral fuels, oils and products (HS code 27) constitute the largest share in Myanmar's export as well as its import basket. In 2018, Myanmar exported and imported USD 3.5 billion and 4 billion worth of mineral fuels, oils and products, respectively (Trademaps database). However, the type of mineral fuel it exports and imports differs.

The exports constitute liquefied natural gas (LNG) (97% of total mineral fuel exports) extracted from the largest reserves in the country, whereas the imports are largely processed petroleum products and oils (95% of total

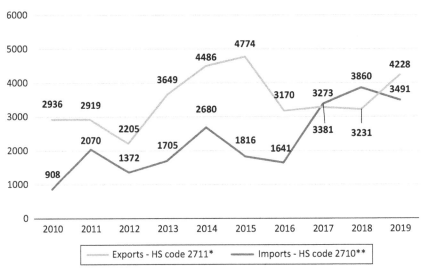

FIGURE 4.2 Myanmar export and import of mineral fuels, oil and products (4 digit HS code) (2010–2018).

Source: Created by the author based on data from ITC Trademap, derived from UN ComtradeNote: *HS Code 2711: Petroleum gas and other gaseous hydrocarbons; **HS Code 2710: Petroleum oils and oils obtained from bituminous minerals (excluding crude)

mineral fuels imports). The export of liquefied natural gas is primarily to China and Thailand. Myanmar currently supplies approximately 6.5 trillion cubic feet of gas to China through the 2,338 km Myanmar–Yunan pipeline. Myanmar is under obligation to continue its gas supply from its large reserves to China for a 30-year period. In 2018, of the total exports of USD 3 billion to Thailand, 73% (USD 2.2 billion) was constituted by natural gas exports (Sinate, Fanai & Bangera, 2019).[1]

In 2018, India's total exports to Myanmar stood at USD 990 million of which petroleum products (excluding crude) constituted 14.8% (USD 147 million). Even with shared borders of 1,643 km, the majority of the trade happens through the sea route (Sinate, Fanai & Bangera, India–Myanmar Trade and Investment: Prospects and Way forward. Working Paper no. 90, 2019). The NER has been experiencing a rise in consumption of POL products, leading to plans for expansion of its existing refining capacity with the rest of India, the benefits of which can also be gained by Myanmar as well.

With the rise in economic activity, Myanmar is also experiencing an increase in energy demand as well. India has seen this as an opportunity to mark its presence by extending its country- and market-level establishments in the energy sector of Myanmar. Between 2009–2018, 80% of the total Indian overseas investment in Myanmar was in the oil and gas sector (Sinate, Fanai & Bangera, India–Myanmar Trade and Investment: Prospects and Way forward. Working Paper no. 90, 2019). Some of the Indian oil and gas companies which have established their presence in Myanmar are provided in Table 4.9.

We can see from Table 4.9 above that both private and public sector Indian companies have their presence in Myanmar. However, even with more than USD 600 million investments, it is still not close to the multi-billion-dollar investments that China has made in the mining and natural gas pipeline projects which are crowding out India's proposals to build energy trade linkages. Numaligarh Refinery Limited (NRL), Assam is currently exporting high speed diesel and high-quality wax to Myanmar via road transport through the Moreh–Tamu border road (Singh, 2017). India had proposed a 7,000 km gas pipeline for export through Myanmar to Bangladesh, the NER and West Bengal which will improve energy connectivity in the NER (Chowdhury, 2017). However, the deal has not materialized as Myanmar's domestic gas supply is already under pressure due to its obligation toward China for gas supply.

9 Comparative Cost of Electricity Generation between the NER and Adjacent Countries

In light of the analysis presented in the previous sections, to explore the trade relations, it is important to undertake the cost analysis of electricity generation among the NER and adjacent countries. Currently, India is having a

TABLE 4.9 Investment by Indian oil and gas companies in Myanmar (2009–2018)

Sr. no.	Particulars	Type of Business	USD million
1.	Production sharing contract between Myanmar Oil and Gas Enterprise and GAIL (India) Ltd. and Silver Wave Energy Pte. Ltd. (Singapore)	Exploration and production of crude oil and natural gas (Block A-7)	47.5
2.	ONGC Videsh Ltd.	Crude oil and natural gas exploration and production (Block AD-2), Yakhine Offshore Deep-Water Area	45
3.	ONGC Videsh Ltd.	Crude oil and natural gas exploration and production (Block AD-3), Yakhine Offshore Deep-Water Area	46
4.	ONGC Videsh Ltd.	Crude oil and natural gas exploration and production (Block AD-9), Yakhine Offshore Deep-Water Area	46
5.	Jubiliant Oil and Gas Pvt. Ltd. and Parami Energy Development (PSC-1)	Crude oil and natural gas exploration and production (Onshore Block I), Hinthada Area Ayeyarwaddy Region	73
6.	ONGC Videsh Ltd. and Machinery and Solutions Co. Ltd. (B-2)	Exploration and production of crude oil and natural gas	49.6
7.	ONGC Videsh Ltd. and Machinery and Solutions Co. Ltd. (EP-3)	Exploration and production of crude oil and natural gas	32.7
8.	Oil India Ltd. and Mercator Petroleum Private Ltd. and Oilmax Energy Private Ltd. and Oil Star Management Services Company Ltd. (YEB)	Exploration and production of crude oil and natural gas	60.5
9.	Oil India Ltd. and Mercator Petroleum Private Ltd. and Oilmax Energy Private Ltd. and Oil Star Management Services Company Ltd. (M4)	Exploration and production of crude oil and natural gas	60.5
10.	Reliance Industries Limited and United National Resources Development Services Co. Ltd. (M-17)	Exploration and production of crude oil and natural gas	116.5
11.	Reliance Industries Limited and United National Resources Development Services Co. Ltd. (M-18)	Exploration and production of crude oil and natural gas	91.5
	Oil and Gas Total		668.7

Source: Sinate, Fanai & Bangera (2019)

TABLE 4.10 Per unit Average Cost of Supplying (ACS) electricity (2019) (in INR/kWh)

Sr. No.	Countries	Average* Cost of Supply (ACS)
1.	China	6.46
2.	Bangladesh	6.20
3.	Myanmar	6.08
4.	NER	6.15
5.	Rest of India**	5.94

Source: GlobalPetrolPrices.com (data accessed in June, 2020); PFC (2020)
*The average has been calculated as the average of ACS for Household and Business Sector. We have not taken into consideration ACS of Bhutan, because Bhutan exports power to India only.
** Rest of India constitutes states other than the states from the NER

comparative advantage in terms of Per Unit Average Cost of Supply (ACS) of electricity. The following shows the per unit average cost of supplying electricity in 2019.

From Table 4.10, it is clear that compared to all the neighboring countries (including the NER), China's ACS is higher, implying that there is a possibility for the NER to capture the electricity trading opportunity in the region due to relatively lower ACS. However, Bangladesh and Myanmar import electricity from China only because of the non-availability of power from India; if India started producing more electricity, it could easily replace China. In order to avail of the opportunity created by the comparative advantage, the electricity distribution segment in the NER also needs to be reformed.

From Table 4.11, it is evident that the financial condition of the State Distribution Utilities is not at all good, with Meghalaya, Manipur, Arunachal Pradesh, Nagaland, Sikkim and Mizoram incurring losses. If the distribution companies are unable to pay their dues to the generating companies, then the power plants remain captive.

Furthermore, in Figure 4.3, we can see that, except Mizoram and Assam, all the other states have AT&C losses, i.e., losses through billing inefficiency of electricity consumption through leakages or theft, higher than the national average of 22%. Thus, distribution companies in the NER need to financially reform in order to assure the smooth flow of electricity from the point of generation to the point of consumption. Cross-border grid integration will be imperative to facilitate the flow of electricity in this case. The physical infrastructure should be developed strategically to ensure smooth flow of electricity at optimal cost (Mukhopadhyay & Prabhu, 2020). When the three important segments of the electricity sector in India, namely generation, distribution and transmission, are operating efficiently together, only then can the comparative advantage of the NER in cross-border trade be explored.

TABLE 4.11 Average Cost of Supply (ACS) and Average Revenue Received (ARR) by State distribution utilities in 2018–2019 (in INR/kWh)

Sr. No.	States	Average Cost of Supply (ACS)	Average Revenue Received (ARR)	Revenue generation(+)/ Loss(−)
1.	Assam	6.53	6.85	0.32
2.	Meghalaya	4.95	4.10	−0.85
3.	Manipur	5.79	5.69	−0.1
4.	Tripura*	4.24	4.30	0.06
5.	Arunachal Pradesh	7.09	2.82	−4.27
6.	Mizoram	7.46	6.28	−1.18
7.	Nagaland	9.74	5.68	−4.06
8.	Sikkim	3.41	3.39	−0.02
9.	All India average	6.09	5.57	−0.52

Source: PFC (2020) and authors' calculations

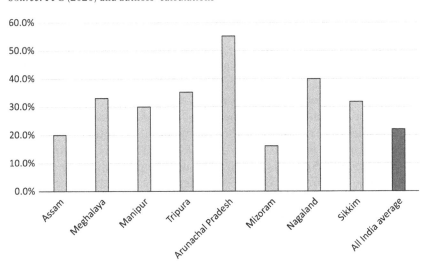

FIGURE 4.3 State-wise aggregate technical and commercial losses (2018–2019) (in percentage).

Source: Created by the author based on data from PFC (2020)

10 Renewable Energy Policy in India: Feasibility Check from a Technical and Financial Point of View

In a significant move, the government also accorded RE status to large Hydro Electric Plants (HEPs) in March 2019, enabling new HEPs to receive concessions and green financing available to RE projects (GOI, 2019). It is aligned with the targets announced in the recently concluded COP26 in Glasgow, where India announced their ambitious aims to (1) achieve 500 GW RE capacity, (2) meet 50% of its energy requirements from non-fossil fuel sources and (3) reduce the carbon intensity of its economy by 45% by 2030 (PIB, 2021).

The choice of the source of power generation has environmental consequences. The power generation in the NER increased by 88% between 2014 and 2019. The CO_2 emissions from the power sector have also increased by 75% during the same period (CEA, 2019a). This is primarily the result of an increase in net power generation from thermal-based power plants in the states of Assam and Tripura. On the other hand, the rest of the six states of the NER are completely dependent on Hydro Electric Plants (HEP) and very few thermal plants which are operational remain captive. This requires close scrutiny of CO_2 emissions from power plants in the NER, Assam and Tripura (Figure 4.4).

From Figure 4.4, it is observed that the trend of absolute CO_2 emissions is rising. As of 2018–2019, the power generation (in GWh) and absolute CO_2 emissions (in tons) was 5,309.7 GWh and 42,64,446 tCO_2 in Assam and 6409.2 GWh and 28,81,665 tCO_2 in Tripura, respectively. All five power plants in Tripura are natural gas-based which, relative to coal, oil and diesel, is a cleaner energy source, hence the CO_2 emissions in Tripura are far lower compared to Assam. If large hydropower potential can substitute even half of the power generated from thermal power plants this will result in a significant reduction in CO_2 emissions as well as simultaneously covering the energy deficit in the region.

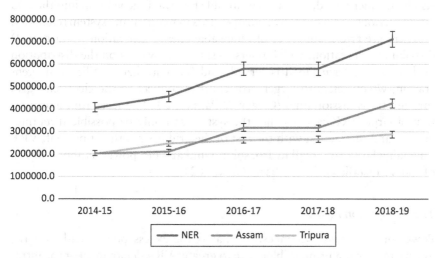

FIGURE 4.4 Absolute CO_2 emissions from power plants in the NER, Assam and Tripura (2014–2019) (in tonnes).

Source: Created by the author based on data from CEA (2019)

Due to various socio-political reasons, the Central and State governments have never attempted to unleash the region's hydropower potential. But now, with a great change in the global geo-political situation and a complete overhaul of India's growth narrative, ensuring energy security in the NER has become a priority for the nation. As a policy directive, improved energy supply can serve a dual objective for the country. First, industrial investment can grow substantially in the NER due to uninterrupted power supply. Second, the NER can tap markets in adjacent countries in South East Asia by exporting an additional supply of power.

Cooperation in the energy sector is always considered to be one of the most important routes to integrate economies with close proximity. Two of the most important trade blocs operational in South and South East Asia (ASEAN and BIMSTEC) have put huge emphasis on collaboration in energy trade since their inception. The BIMSTEC Grid Interconnection Program was established with the aim of expanding energy trade among Member States, and has accelerating development of new hydropower projects as one of its objectives (GOI, 2018). In light of growing energy and trade importance, the Government of India is exploring trade relations for securing power sector development in a climate-resilient manner.

11 Generating More RE: Is This a Prudent Policy for India? Examining from Technical Aspect

11.1 Grid Stability

As already mentioned, integrating variable renewable generation into the grid poses a serious challenge to system operators in maintaining system reliability.

There are two main types of renewable energy generation resources: (1) distributed generation, which refers to small renewables on the distribution grid where electricity load is served; and (2) centralized, utility-scale generation, which refers to larger projects that connect to the electricity grid through transmission lines (Cleary & Palmer, 2020). Successful incorporation of large VRE capacity into the system can only be possible if its integration and interaction with the rest of the grid is smooth. At present, VRE (including large and small hydro, solar, wind and bio-power) is about 36.7% of India's installed capacity base (CEA, 2021).

11.2 Congestion Issues

Power or transmission congestion that is the excess power load over the given capacity is a primary blockage to greater VRE deployment at the intra-state as well as inter-state level during the period of power surplus. A balance between VRE and conventional energy sources is to be achieved to reduce congestion. Certain critical areas for analysis such as the location of transmission congestion, the availability of resources to meet grid surges and system balancing costs should have high priority for further investigation as they will have a bearing on grid stability that will take time to deliver (Shakti Foundation and GE India Exports Pvt. Ltd., 2018).

11.3 Construction of Transmission Infrastructure

As discussed earlier, the interstate and inter-regional transmission infrastructure has already been developed with the five synchronously connected regional grids. However, new transmission corridors would be required for evacuating green energy from RE-rich states such as Tamil Nadu, Gujarat and Rajasthan. Transmission planners now recognize that in view of the short gestation period of RE plants, the transmission segment has to lead generation and would require upfront investment. Transmission corridors have been identified for construction in the next five years through the established process of coordinated transmission planning (Shakti Foundation and GE India Exports Pvt. Ltd., 2018).

11.4 Efficient Forecasting

Generation from VRE Sources depends on nature, i.e., rainfall, wind velocity and sunshine. The variability of VRE power can be addressed through advanced forecasting techniques, which are still evolving in India. The grid operators must have a sense of the RE supply potential and the expected demand from the grid at any given point in time. This task is especially challenging due to the intermittent nature of RE sources and their variation in size and locations across the power grid (Katz & Cochran, 2015). As the share of VRE capacity on the grid grows, these issues become increasingly important to understand.

11.5 Energy Hierarchy

One of the effective measures to ensure grid stability is to ascertain the energy hierarchy.

The hierarchical interaction depends to a great extent on the number of energy options available to the grid operator. Swapping from one to another option in accordance with the electricity demand and supply is possible only when a number of options are available. The NER of India produces almost all types of energy as shown in Table 4.13. Here renewable energy sources represent all kinds of VRE: solar, wind, biomass, etc. If we include hydro in this, the ratio of available energy options in the NER and at the All-India level is shown below in Table 4.12.

TABLE 4.12 Ratio of non-VRE to VRE with and without hydro energy installed capacity (as of 31.02.2021)

Sr. no.		NER	All India
1.	Non-VRE:VRE ratio (with hydro)	1:0.89	1:0.59
2.	Non-VRE:VRE ratio (without hydro)	1:0.14	1:0.39

Source: CEA (2019); authors' calculations

We can see from Table 4.12 that with the inclusion of hydro, the disparity between VRE and non-VRE energy scenarios is lower for the NER as well as the All-India level. Without hydro, the disparity between the two types of energy sources is far higher in the NER. This shows the huge potential of hydro in the region. However, due to the already inherent instability in the energy ecosystem as explained above, the integration of electricity generation from hydropower is a very tedious process and technically challenging. The financial aspect of this issue is discussed in the following sub-section.

12 Examining from the Financial Aspect

12.1 Cost Implications to Distribution Companies (DisComs)

Almost all of India's electricity distribution companies (DisComs) are publicly owned and are incurring financial losses leading to deferring payments, renegotiating power purchase agreements (PPAs) or avoiding signing new PPAs altogether. The rapid growth in VRE has ramifications on the financial preparedness of distribution companies, consumer tariffs and incumbent power-generating companies, especially coal-fired thermal power plants. Integrating renewables on the power grid with limited impact on tariffs, low curtailment rates and ensuring grid stability demands further techno-economic study of trade-offs, instruments and risk diversification. The top-down RE targets of 175 GW by 2022 and 500 GW by 2030 need to be determined in consultation with the state governments, as the largely state-owned DisComs are to be financially prepared and made viable in order to meet the higher integration of VRE. Since the constitution of India places electricity in the concurrent list, that is, on which both the central and state governments have jurisdiction, the state has to pay the bulk of the cost of grid integration. Spreading the costs of integration across the country must ensure improvements to existing institutions like the Renewable Energy Certificate (REC) mechanism and augmenting the transmission infrastructure. Even if RPOs are spread across all states equitably, these would not automatically cover grid integration costs unless it is purposely designed so. However, it has been found that expanding markets with larger balancing areas and with more flexible operations helps to cover costs significantly

TABLE 4.13 Region-wise and State-wise increase in installed generating capacity of electricity (utilities) in India as of 28.02.2020 and 28.02.2021 (in MW)

States	Hydro		Thermal		Nuclear		RES*		Total		Growth rate (in %)
	31.02.20	31.02.21	31.02.20	31.02.21	31.02.20	31.02.21	31.02.20	31.02.21	31.02.20	31.02.21	
Arunachal Pradesh	158.05	544.55	83.87	83.87	0	0	136.72	136.72	378.64	765.14	102.1%
Assam	489.58	522.08	1,191.77	1,192.92	0	0	75.34	79.1	1,756.69	1,794.1	2.1%
Manipur	102.34	95.34	154.67	154.67	0	0	10.61	11.81	267.62	261.82	−2.2%
Meghalaya	401.77	409.27	160.31	160.31	0	0	46.45	46.45	608.53	616.03	1.2%
Mizoram	102.44	97.94	71.51	71.51	0	0	37.99	38	211.94	207.45	−2.1%
Nagaland	61.83	66.33	81.03	81.03	0	0	31.67	31.67	174.53	179.03	2.6%
Tripura	75.99	68.49	662.55	662.55	0	0	25.42	25.42	763.96	756.46	−1.0%
Central Sector NER (unallocated)	185	140	176.12	176.12	0	0	0	0	361.12	316.12	−12.5%
Sub-Total (NER)	1,577	1,944	2,581.83	2,582.98	0	0	364.2	369.17	4,523.03	4,896.15	8.2%
Total All India	45,699.22	46,209.22	230,189.57	233,170.72	6,780	6,780	86,759.19	92,970.48	369,427.97	379,130.41	2.6%

Source: CEA (2020 & 2021)

Note: RES includes solar, wind, bio-power and small hydro

(Tongia, Harish & Walawalkar, 2018). The difficulties of integrating RE into India's power grid will worsen as RE's share of generation increases, causing disproportionate strain on states rich in RE resources.

12.2 Cost of Grid Integration

VRE has not yet developed to the extent that it can compete with or even be a substitute for coal-fired generation, which remains the dominant source of power in India. Low VRE costs have given birth to the new concept of "grid parity" – an imaginary stage when VRE will push coal off the Indian grid. This stage appears to be out of reach in the near future, at least until the whole cost of integrating VRE into the grid is recovered. The best performing RE systems, with due incorporation of cost of integration, are found to be competitive only with the most expensive new coal projects, but not with existing coal plants. RE integration would be easier across larger balancing areas within the grid, but that approach would require substantial green financing in long-distance RE-centric transmission and energy storage systems, which have been limited so far.

Looking to the future, growing RE's share of generation will require governmental support in the form of policy instruments such as subsidies or tax rebates to reduce the cost of grid integration. New market incentives are needed to create the right types of supply based on location, seasonal or daily availability and ramping capabilities. Special emphasis is to be given to revamping the state controlled DisComs, which is the vulnerable segment in the existing system. The highest paying commercial and industrial customers are among the biggest investors in rooftop solar resources. An even bigger push toward RE by these important customers could accelerate the downward spiral of DisCom finances. Holistic policies will accelerate the transition.

13 Cross-Border Power Trade and Regional Grid Integration: A Possible Solution

Since integrating RE with the NER power grid poses some serious challenges to the Indian energy ecosystem, we should simultaneously look for some alternatives to utilize generated RE from the NER. Cross-border power trade and integration with a regional grid could be one of the feasible solutions.

Addressing the growing demand for energy in the ASEAN region, the 32nd ASEAN Ministers on Energy Meeting (AMEM) held on 23 September 2014 in Vientiane, Lao PDR, endorsed the theme of the new ASEAN Plan of Action for Energy Cooperation (APAEC) 2016–2025. The key initiatives are given below:

i. Embarking on multilateral electricity trading to accelerate the realization of the ASEAN power grid (APG).

ii. Enhancing gas connectivity by expanding the focus of the Trans-ASEAN Gas Pipeline (TAGP) to include liquefied natural gas regasification terminals as well as promoting clean coal technologies.

iii. Strategies to achieve higher aspirational targets to improve energy efficiency and increase the uptake of RE sources, in addition to building capabilities on nuclear energy.

iv. Plans to broaden and deepen collaboration with ASEAN's Dialogue Partners (DPs), International Organizations (IOs), academic institutions and the business sector will be stepped up to benefit from their expertise and enhance capacity-building in the region (ASEAN Centre for Energy, 2015).

The ASEAN power grid was an initiative to construct a regional power interconnection eventually expanding to a completely integrated South East Asia power grid system. The APG project is expected to enhance electricity trade across borders, and in order to take advantage of this and engage in cross-border power trade India should try to get connected with the APG through Myanmar's transmission network.

14 Market Preferences and Regulatory Choices for the ASEAN and North East States

For energy demand, the NER states buy power either from private power producers or get supply from the NER's National Load Despatch Centres (NERLDC); for supplying, its market consists of local demand and other power deficit states through a network of five LDCs. The National Load Despatch Centre (NLDC) has, however, the option to engage in cross-border electricity trade.

14.1 Regulatory Framework for Domestic and Cross-Border Trade

Since the economic liberalization policies of 1991, states across India were unbundled and created separate generators, transmission companies and distribution companies. Each state is now ultimately responsible for buying sufficient power to meet consumer demand, and also runs its own Load Despatch Centre (LDC) to choose which suppliers to call at what time to meet instantaneous demand. State LDCs take care of grid stability and try to run the system at the lowest cost. State and regional LDCs coordinate closely for management of inter-state flows of power (Tongia, Harish & Walawalkar, 2018).

In case of default in payment or non-maintenance of a payment security mechanism as per agreement, Regional Load Despatch Centres (RLDCs) shall prepare implementation plan for regulation of power supply based on request of generating company or transmission company, as per CERC (Regulation of Power Supply) Regulations, 2010.

The Power System Operation Corporation Limited (POSOCO) is responsible for regulation and inter-regional and transnational exchange of power supply. POSOCO consists of five RLDCs and the NLDC.

The five RLDCs oversee the interstate transmission, as is shown in Figure 4.5 in the appendix.

Along with being one of the largest electricity producers in Asia as well as across the world, there are still several shortcomings in India's electricity system such as shortages in meeting peak times (3.2%) as well as the energy shortage (2.1%) witnessed during FY2015–2016. In spite of the overall shortage, the inherent diversity in demand of various states and regions in the country results in periods of seasonal surplus in one state with complementary periods of deficit in another. PTC India Ltd. (formerly known as Power Trading Corporation of India Limited) was incorporated in 1999 to undertake trading of power to achieve economic efficiency, security of supply and to develop a vibrant power market in the country. Therefore, PTC has a tri-fold mandate:

1. To optimally utilize the existing resources to develop a fully fledged efficient and competitive power market.
2. To attract private investment in the Indian power sector.
3. To encourage trade of power with neighboring countries.

The pioneering service of the Company led to recognition of "Power Trading" as a distinct licensed activity in the Electricity Act 2003 (PTC India, 2019).

In order to facilitate the cross-border energy trade, the Ministry of Power, Government of India has published the guidelines to cross-border trade of electricity, 2018, which succeeded the 2016 guidelines (GOI, 2018; Batra, 2020). Some of the important features of the new guidelines which are distinct from the previous guidelines are given below:

i. Unlike the 2016 guidelines which allowed only bilateral trading, in the new guidelines, trilateral trading of electricity, i.e., trading between two countries not sharing a common physical border and passes through an intermediate country is allowed.
ii. Other countries can also trade on the power exchange in India.

Under the Cross-Border trade of Electricity Regulations (2019), the sale and purchase of power between India and the neighboring countries will be allowed through mutual agreements between the local entities of respective countries through bilateral agreements or bidding route (CERC, 2019).

The new regulations require establish the following institutional framework:

a) A designated authority appointed by the power ministry for facilitating the trade.
b) A transmission planning agency – to facilitate cross-border trade.
c) A nodal agency for settling charges.
d) System operator.
e) Central transmission utility – responsible for granting long-term and medium term access to the grid.

15 Enhancing Cross-Border Electricity Trade through Load Despatch Centres (LDCs)

Under the new regulations, the National Load Despatch Centres will act as the system operator, i.e., it will undertake the responsibility of transporting electric power between neighboring countries (Verma, 2019). NLDCs supervise and control the inter-regional links through the national grid in order to maintain grid stability and also ensure grid security. Other major responsibilities of NLDCs are regulation of power supply, ensuring synchronous operation of national grid and supervision of Regional Load Despatch Centres (POSOCO, 2017). The apex body monitoring the NLDCs is the Power System Operation Corporation Limited which undertakes the responsibility of supervision and operational requirement of all NLDCs. Under the given regulatory framework, the exchange of power between India, Myanmar and other ASEAN members can be extended through hierarchical interactions at the country level in which POSOCO can undertake the major responsibility of ensuring seamless cross-border electricity trade from LDCs in the NER to neighboring countries. Companies such as North Eastern Electric Power Corporation (NEEPCO) which is one of the largest power-generating companies can play a significant role in this aspect.

16 Common Grid Code

As CERC has authority only within the borders of India and not in any of its neighboring partner countries, a common grid code is proposed for trading of electricity across the border (Batra, 2020). This was conceptualized by CERC. Some of the features of the common grid code are given below.

i. It is a structure for preparing a common ground for electricity trade and is implemented at different platforms – governments, regulators, planning, transmission utilities, system operators and accounts settlement.
ii. It ensures equitable participation of all countries.

iii. It provides a uniform transmission planning criterion.
iv. It is applicable for synchronous (AC) and asynchronous connections (DC).
v. It constitutes four parts – the connection code, operating code, scheduling and dispatch code and administration of the grid code.
vi. It proposes installation of data acquisition systems, disturbance recorders and sequences of events recorders oat the interconnection points and other significant points, protection, coordination in the partnering countries.

17 Conclusion and Policy Implications for the NER and ASEAN for Trade and Connectivity

Given the geo-strategic and geo-economic significance of the South China Sea playing a significant role in China's influence on the ASEAN countries in the region, it is time for India to become proactive in establishing regional cooperation and economic integration in South East Asia. For this purpose, the NER offers a gateway to ASEAN countries in the energy sector through hierarchical interactions at the country-level and company-level as well.

Our study shows that the large hydropower potential in the NER could play a pivotal role in cross-border electricity trade with Myanmar as well as other ASEAN member countries. However, as hydropower generation is intermittent and variable in nature it cannot be sustainably integrated into the existing electricity grid infrastructure due to the technical and financial vacuum that exists for expediting its grid integration. This lacuna can be overcome through hierarchical interactions at the country-level such as mutually agreed market preferences and regulatory choices to optimize and upgrade the grid infrastructure, as well as at company-level by providing energy market forecasting in order to reduce the magnitude of risk arising from uncertainty in energy demand–supply requirement. By ensuring grid stability and security, the large hydropower potential can be realized, which will provide further impetus for cross-border power trade with ASEAN countries through Myanmar's transmission network of Myanmar. This would allow India to integrate with the APG, as well as establish the ambitious "One Sun, One World, One Grid" project in future.

The findings of our study have resulted in a few interesting policy suggestions:

1. Even though the NER is considered a gateway to ASEAN countries, it cannot be treated in isolation from the rest of India. The NER's integration with the rest of India is of significant importance as it would be a catalyst for cross-border energy trade with Myanmar, and eventually further extending to the rest of the ASEAN members. Strengthening

the NER's physical and energy infrastructure with the rest of India will further enhance the entire nation's trade with ASEAN members. The achievement of "One Nation, One Grid" is a perfect example of providing higher impetus for energy security in the NER and furthering the prospect of electricity trade with Myanmar.

2. A fixed proportion of the budget allotted to develop the power sector must be allocated for the hydro sector separately. Since the gestation period for most of the hydropower plants exceeds the political tenure of five years in India, there is a tendency to spend for thermal power, where outcomes are more immediate.

3. As water and water supply are state subjects, the construction of Hydroelectric plants (HEPs) can be delayed due to the conflicting interests of affected states due to the expected disruption of water flow – the Subansiri HEP is a prime example of this. Hydropower projects should be brought on the concurrent list such that there will be proper consultation and feasibility studies after considering the inputs of all stakeholders involved and formulation of a uniform policy by the central government can initiate this process faster. Most of the HEPs are located in difficult and inaccessible terrain. They require the development of physical infrastructure for project implementation. Roads and bridges also provide higher opportunities for the development of neighbouring areas. At present, the government largely provides funds for the construction of HEPs, especially in hilly and difficult terrain, but the process to grant financial support needs to be streamlined.

4. Developing countries should never go for long-term trade contracts with other countries. In this era of ever-changing growth dynamics, a contract of 20/30 years is too long. Situations may change to such an extent that these contracts become a major impediment to the growth of the country.

5. At present, the regulatory clearances from different authorities and government bodies are required for HEPs with capital expenditure above ₹1000 crore which leads to a long gestation period for the project to get sanctioned. Processes must be revisited to reduce the time taken.

6. The regional integration should be very comprehensive, taking into consideration all other sectors. For example, if the transport sector lags behind, all other sectors will surely be affected. This happens very fast in a highly interdependent industry structure.

7. The National Load Despatch Centres should operate as mediating and monitoring entities for cross-border trade between the NER and ASEAN members.

8. Hydropower Purchase Obligation (HPO) should be implemented in the upcoming "Electricity (Amendment) Bill, 2020".

9. There are few renewable energy resources available in the NER which are DC-based, such as photovoltaic (PV) fuel cells (FC), etc. Hence, for grid integration and power trade, the NER should put more focus on these DC-based renewable energy resources.

10. Some researchers have suggested an alternative hierarchy to design a more efficient electricity ecosystem (Khan et al., 2015). The DC–DC converters are more suitable for energy storage purposes and grid integration on the distribution side. The government should venture into similar hierarchical interactions devised by researchers of the country's esteemed engineering institutions like IITs to find better options. More funding should be provided by the government to these academic institutions for this purpose.

11. A micro grid has a pivotal role to play in the transformation of the existing power grid to a sophisticated smart grid in the future. In the NER, several nano grids serving in geographically closed locations can be combined to form a community micro grid (Paudel & Beng, 2018). The Government of India should incentivize the adoption of nano grids in the NER which is attracted for the DisComs as well as the local consumers. Such policy interventions can act as a catalyst in increasing adoption of RE generation and energy storage systems in the residential sectors.

12. Along with cross-border power trade with Myanmar and extending it to the other ASEAN members, hierarchical interactions in the oil and gas sector at the country and company level between the NER of India and Myanmar will also further strengthen energy cooperation in the region and lead to a positive multiplier effect for both sides, and at the same time counter China's monopolistic influence in the energy sector in the region. With the oil refinery in Numaligarh, Assam is already playing a significant role in exporting high speed diesel to Myanmar. Further increasing the intensity of export through the expansion of pipelines should be an area of further research.

13. In addition to the above policy implications, the environmental impacts must be considered within the decision-making framework. As a result of the NER's tremendous hydro potential, a phasing-out process should be initiated for coal-, oil-, diesel- and gas-fired thermal power plants in the states of Assam and Tripura. The power plants in the other six states in the NER are predominantly HEPs. With more than 90% of hydropower potential yet to be explored in the region, power generated through HEPs can be supplied to Assam and Tripura as well as meeting the rising electrical energy requirements in the entire region. Even though this will be a long-term process, the transition toward cleaner energy sources will also be a step toward Sustainable Development Goal 7: providing affordable and clean energy for all, not only in the NER but

also in Myanmar and the other ASEAN member states through cross-border energy trade from HEPs.

Note

1 www.trademap.org/Index.aspx?AspxAutoDetectCookieSupport=1.

References

Ahmad, S., & Tahar, R. M. (2014). Selection of renewable energy sources for sustainable development of electricity generation system using analytic hierarchy process: A case of Malaysia. *Renewable Energy, 63*, 458–466. http://doi.org/10.1016/j.renene.2013.10.001.

ASEAN Centre for Energy. (2015). *ASEAN plan of action for energy cooperation (APAEC) 2016–2025.* Jakarta, Indonesia: ASEAN Centre for Energy. One Community for Sustainable Energy. Retrieved from https://globalabc.org/sites/default/files/2020-04/ASEAN%20Plan%20Of%20Action%20For%20Energy%20Cooperation%20%28apaec%29.pdf.

Batra, P. (2020). Increasing connectivity. *PowerLine.* Retrieved from https://powerline.net.in/2020/01/27/increasing-connectivity/.

Bhaskar, U. (2020). CCEA hikes budget to improve Northeast's electricity transmission, distribution. *Livemint.* Retrieved from https://www.livemint.com/news/india/ccea-hikes-budget-to-improve-northeast-s-electricity-transmission-architecture-11608116469590.html.

Bhaskar, U., Raj, A., & Ramanathan, A. (2014). India is now one nation, one grid. *Livemint.* Retrieved from https://www.livemint.com/Politics/jIOljqqvinQqqngk7BYLZP/Southern-transmission-line-connected-to-National-Grid.html.

Bose, P. R. (2018). Construction of India-Bangladesh oil pipeline to begin this month. *The Hindu Business Line.* Retrieved from https://www.thehindubusinessline.com/news/construction-of-india-bangladesh-oil-pipeline-to-begin-this-month/article24835076.ece#.

CEA. (2019a). Baseline carbon dioxide emission database version 15.0. Central Electricity Authority, Government of India. Retrieved May 2020, from http://www.cea.nic.in/tpeandce.html.

CEA. (2019b). *National electricity plan (volume II) transmission.* New Delhi, India: Central Electricity Authority, Ministry of Power Government of India. Retrieved August 2020, from https://powermin.gov.in/sites/default/files/uploads/NEP-Trans1.pdf.

CEA. (2020a). *All India installed capacity (in MW) of power stations (As on 29.02.2020).* New Delhi, India: Central Electricity Authority, Ministry of Power, Government of India. Retrieved March 2021, from https://cea.nic.in/wp-content/uploads/installed/2020/06/installed_capacity-02.pdf.

CEA. (2020b). *Load generation balance report 2020–21.* New Delhi, India: Central Electricity Authority, Ministry of Power, Government of India. Retrieved January 2021, from https://cea.nic.in/old/reports/annual/lgbr/lgbr-2020.pdf.

CEA. (2021). *All India installed capacity (in MW) of Power stations (As on 28.02.2021)*. New Delhi, India: Central Electricity Authority, Ministry of Power, Government of India. Retrieved March 2021, from https://cea.nic.in/wp-content /uploads/installed/2021/02/installed_capacity_02.pdf.

CERC. (2019). *Central electricity regulatory commission (cross border trade of electricity) regulations, 2019*. New Delhi, India: Central Electricity Regulatory Commission, Ministry of Power, Government of India. Retrieved from http:// www.cercind.gov.in/2019/regulation/CBTE-Regulations2019.pdf.

Chaudhury, D. R. (2020). *Myanmar continues pushback against BRI: Chinese Eco Development Zone faces turbulence. The Economic Times*. Retrieved from https://economictimes.indiatimes.com/news/international/world-news/ myanmar-continues-pushback-against-bri-chinese-eco-development-zone-faces -turbulence/articleshow/77364893.cms.

Chowdhury, J. R. (2017). Tri-nation pipeline envisaged. *The Telegraph*. Retrieved from https://www.telegraphindia.com/business/tri-nation-pipeline-envisaged/cid /1472565.

Cleary, K., & Palmer, K. (2020). *Renewables 101: Integrating renewable energy resources into the grid*. Resources for the Future. Retrieved from https://www.rff .org/publications/explainers/renewables-101-integrating-renewables/.

Das, R. U. (2016). *Enhancing India-myanmar border trade: Policy and implentation measures*. New Delhi, India: Department of Commerce, Ministry of Commerce, Government of India. Retrieved from https://commerce.gov.in/writereaddata /uploadedfile/MOC_636045268163813180_Final%20Enhancing_India_ Myanmar_Border_Trade_Report.pdf.

DGH, & CRISIL (2015). Hydrocarbon vision 2030 for Northeast India. Directoral General of Hydrocarbons and Crisil Infrastructue Advisory. Retrieved May 2020, from http://www.megplanning.gov.in/circular/Hydrogen%20Vision %20for%20NER%20-%20Draft%20for%20Consultation%20.pdf.

Energetica India. (2020). *India Will add hydropower capacity worth 70,000 MW by 2030: Joint secretary, MNRE*. Energetica India. Retrieved from https://www .energetica-india.net/news/india-will-add-hydropower-capacity-worth-70000 -mw-by-2030-joint-secretary-mnre#:~:text=India%20has%20an%20estimated %20hydropower,the%20state%20power%20distribution%20companies.

ET Energyworld. (2020). *India exported 8 BUs of electricity in Apr-Jan period of current financial year*. ET Energyworld. Retrieved from https://energy .economictimes.indiatimes.com/news/power/india-exported-8-bus-of-electricity -in-apr-jan-period-of-current-financial-year/74732940.

GOI. (2018). *Memorandum of understanding for establishment of the BIMSTEC Grid interconnection*. Ministry of Power, Government of India. Retrieved from https://powermin.nic.in/sites/default/files/uploads/BIMSTEC_MoU.pdf.

GOI. (2019). *Standing committee on energy (2018–2019)*. New Delhi: Lok Sabha Secretariat, Government of India. Retrieved February 2021, from http://164.100.47.193/lsscommittee/Energy/16_Energy_43.pdf.

GOI. (2020). *India-myanmar joint statement during the state visit of the president of myanmar to India (February 26–29, 2020)*. Ministry of External Affairs, Government of India. Retrieved from https://mea.gov.in/bilateral-documents .htm?dtl/32435/IndiaMyanmar+Joint+Statement+during+the+State+Visit+of+the +President+of+Myanmar+to+India+February+2629+2020.

Embassy of India, Yangon. (2019). *Nilateral economic & commercial relations.* Embassy of India at Yangon. Retrieved from https://embassyofindiayangon.gov .in/pdf/menu/BILATERAL_May3-19.pdf.

Government of India. 2015 EXP AND PROD - NORTHEAST | Ministry of Petroleum and Natural Gas | Government of India - Ministry of Petroleum And Natural Gas (mopng.gov.in).

Govt. of Myanmar. (2021). *Border trade data.* Ministry of Commerce, The Republic of the Union of Myanmar. Retrieved from https://www.commerce.gov.mm/en/ dobt/border-trade-data.

Ilic, D., Da Silva, P.-G., Karnouskos, S., & Griesemer, M. (2012). An energy market for trading electricity in smart grid neighbourhoods. 6th IEEE International Conference on Digital Ecosystems and Technologies (DEST) (pp. 1–6). Campione d'Italia, Italy. https://doi.org/10.1109/DEST.2012.6227918.

PTC India. (2019). *The concept of PTC.* Power Trading Corporation of India Limited. Retrieved from https://www.ptcindia.com/about-us/the-concept-of-ptc/.

Joshi, A. (2020). *North India registered steepest fall in power demand during April 5 lights-out event.* ET Energyworld. Retrieved from https://energy.economictimes .indiatimes.com/news/power/north-india-registered-steepest-fall-in-power -demand-during-april-5-lights-out-event/75039333.

Katz, J., & Cochran, J. (2015). Integrating variable renewable energy into the grid: Key issues. *Greening the grid.* Retrieved from https://greeningthegrid.org /resources/factsheets/integrating-variable-renewable-energy-into-the-grid-key -issues#:~:text=Variable%20renewable%20energy%20(VRE)%3A,and%20 some%20hydropower%20generation%20technologies.

Kesavan, S. (2019). Seamless exchange. *Powerline.* Retrieved from https://powerline .net.in/2019/04/01/seamless-exchange/.

Khan, M., Karimoy, W., & Saeed, M. (2015). New proposed hierarchy for renewable energy generation to distribution grid. 2015 International Conference on Emerging Technologies (ICET). Institute of Electrical and Electronics Engineers. https://doi.org/10.1109/ICET.2015.7389187.

Kumar, A. (2018). *Electricity trade in South Asian could grow up to 60,000 Mw through 2045.* ET. Energyworld.com.Retrieved from https://energy .economictimes.indiatimes.com/news/power/electricity-trade-in-south-asian -could-grow-up-to-60000-mw-through-2045/66477793#:~:text=India's%20p ower%20trade%20with%20its,for%20trade%20with%20Sri%20Lanka.

Kumar, P., & Krishnan, D. S. (2020). *Lights out at 9: Lessons from the successful experiment with India's energy infrastructure.* News18 Opinion. Retrieved from https://www.news18.com/news/opinion/lights-out-at-9-lessons-from-the -successful-experiment-with-indias-energy-infrastructure-2578931.html.

La Rovere, E. L., Soares, J. B., Oliveira, L. B., & Lauria, T. (2010). Sustainable expansion of electricity sector: Sustainability indicators as an instrument to support decision making. *Renewable and Sustainable Energy Reviews*, 14(1), 422–429. https://doi.org/10.1016/j.rser.2009.07.033.

Mathiesen, B. V., Lund, H., Connolly, D., Wenzel, H., Ostegraard, P. A., Moller, B.,... Hvelplund, F. K. (2015). Smart Energy Systems for coherent 100% renewable energy and transport solutions. *Applied Energy*, 139–154. https://doi .org/10.1016/j.apenergy.2015.01.075.

MNRE. (2020). *Request for proposals for Developing a long-term vision, implementation plan, road map and institutional framework for implementing 'One Sun One world One Grid'*. New Delhi: Ministry of New and Renewable Energy, Government of India. Retrieved from https://mnre.gov.in/img/documents/uploads/file_f-1590573144563.pdf.

Mukhopadhyay, K., & Prabhu, V. S. (2020). Debacle of the power policy in India: Generation, transmission, distribution and regulation. IAEE Energy Forum, Second Quarter 2020, 31–35. Retrieved from http://india.iaee.org/~iaeeorg/debacle-of-the-power-policy-in-india-generation-transmission-distribution-and-regulation/.

NEEPCO. (2017). *Power potential in the north eastern region*. North Eastern Electric Power Corporation Limited. Retrieved from https://neepco.co.in/projects/power-potential.

NERPC. (2018). *Annual report 2017–18*. Shillong: North Eastern Regional Power Committee, Central Electricity Authority, Ministry of Power, Government of India. Retrieved from http://www.nerpc.nic.in/Reports/Annual%20Reports/Annual%20Report%202017-18_Final.pdf.

Paudel, A., & Beng, G. H. (2018). A hierarchical peer-to-peer energy trading in community microgrid distribution systems. IEEE Power and Energy Society General Meeting (PESGM) (pp. 1–5). Portland: Institute of Electrical and Electronics Engineers. https://doi.org/10.1109/PESGM.2018.8586168.

PFC. (2020). *Report on performance of state power utilities 2018–19*. New Delhi, India: Power Finance Corporation, Government of India. Retrieved from https://www.pfcindia.com/DocumentRepository/ckfinder/files/Operations/Performance_Reports_of_State_Power_Utilities/Report_on_Performance_of_State_Power_Utilities_2018_19.pdf.

PFC. (2020). *Report on performance of state power utilities 2018–19*. New Delhi, India: Power Finance Corporation, Government of India. Retrieved March 2021, from https://www.pfcindia.com/DocumentRepository/ckfinder/files/Operations/Performance_Reports_of_State_Power_Utilities/Report_on_Performance_of_State_Power_Utilities_2018_19.pdf.

PIB. (2015). *Year end review - Solar power target reset to one lakh MW*. Press Information Bureau, Government of India. Retrieved from https://pib.gov.in/newsite/PrintRelease.aspx?relid=133220.

PIB. (2021). *National statement by Prime Minister Shri Narendra Modi at COP26 summit in Glasgow*. Press Information Bureau, Government of India. Retrieved from https://pib.gov.in/PressReleasePage.aspx?PRID=1768712.

Pillai, A. V., & Prasai, S. (2018). *The price of power: The political economy of electricity trade and hydropower in eastern South Asia*. New Delhi, India: The Asia Foundation. Retrieved August 2020, from https://asiafoundation.org/wp-content/uploads/2018/07/The-Price-of-Power-The-Political-Economy-of-Electricity-Trade-and-Hydropower-in-Eastern-South-Asia.pdf.

POSOCO. (2017). *National Load Dispatch Centre*. Power System Operation Corporation Limited. Retrieved from https://posoco.in/about-us/.

POWERGRID. (2020). *Smart grid*. Power Grid Corporation of India Limited. Retrieved from https://www.powergridindia.com/smart-grid.

Raghavan, P. (2019). Infrastructure, Budger 2019: 'One Nation, One Grid' to make power cost for states affordable. *Indian Express*. Retrieved from https://

indianexpress.com/article/business/budget/union-budget-2019-one-nation-one
-grid-to-make-power-cost-for-states-affordable-5817887/.

Rahman, M. A. (2020). Bangladesh starts exporting LPG to Indian state of Tripura. *The Financial Express*. Retrieved from https://thefinancialexpress.com.bd/trade/ bangladesh-starts-exporting-lpg-to-indian-state-of-tripura-1578890659.

Ray Chaudhury, A., & Basu, P. (2015). Proximity to connectivity: India and its eastern and southeastern neighbours PART 2. India-myanmar connectivity: Possibilities and challenges. Observer Research Foundation. Retrieved from https://www.orfonline.org/wp-content/uploads/2015/12/IndiaMyanmar.pdf.

Ray Chaudhury, A., & Basu, P. (2015). *Proximity to connectivity: India and its eastern and southeastern neighbours PART 2. India-myanmar connectivity: Possibilities and challenges.* Observer Research Foundation.

Roberts, E. B. (1978). *Managerial applications of system dynamics.* Cambridge: Pegasus Communications. Retrieved from https://systemdynamics.org/product/ managerial-applications-of-system-dynamics-1999-originally-published-in-1978/.

Shakti Foundation and, GE India Exports Pvt. Ltd. (2018). *Integrating renewable energy modelling with power sector planning for India.* Shakti Sustainable Energy Doundation. Retrieved from https://shaktifoundation.in/wp-content/ uploads/2018/01/Study-Report_Integrating-RE-modelling.pdf.

Sinate, D., Fanai, V., & Bangera, S. (2019). *India-Myanmar trade and investment: Prospects and Way forward.* Working Paper no. 90. Export import (EXIM) Bank of India. Retrieved September 2020, from https://www.eximbankindia.in/Assets /Dynamic/PDF/Publication-Resources/ResearchPapers/110file.pdf.

Singh, B. (2017). Numaligarh refinery limited commenced export of high speed diesel to Myanmar. *The Economic Times*. Retrieved from https://m.economictimes .com/industry/energy/oil-gas/numaligarh-refinery-limited-commenced-export-of -high-speed-diesel-to-myanmar/articleshow/60365761.cms.

Stranzali, E., & Aravossis, K. (2016). Decision making in renewable energy investments: A review. *Renewable and Sustainable Energy Reviews*, *55*, 885– 898. https://www.sciencedirect.com/science/article/abs/pii/S1364032115012733.

Taneja, N., Naing, T. H., Joshi, S., Singh, T. B., Bimal, S., Garg, S., … Sharma, M. (2019). *India myanmar border trade.* Working Paper 378. Indian Council for Research on International Economic Relations (ICRIER). Retrieved from http:// icrier.org/pdf/Working_Paper_378.pdf.

Tongia, R., Harish, S., & Walawalkar, R. (2018). *Integrating Renewable Energy into India's Grid - Harder than it looks.* Brookings India. Retrieved from https:// www.brookings.edu/wp-content/uploads/2018/11/Complexities-of-Integrating -RE-into-Indias-grid2.pdf.

Tziogas, C., Georgiadis, P., & Papadopoulos, A. (2019). Fostering the transition to sustainable electricity systems: A hierarchical analysis framework. *Journal of Cleaner Production*, *206*, 51–65. https://doi.org/10.1016/j.jclepro.2018.09.117.

Verbong, G. P., & Geels, F. W. (2010). Exploring sustainability transitions in the electricity sector with socio-technical pathways. *Technological Forecasting and Social Change*, *77*(8), 1214–1221. https://doi.org/10.1016/j.techfore.2010.04 .008.

Verma, A. (2019*). CERC Issues New Regulation for cross border trade of electricity.* SAUR Energy International. Retrieved from https://www.saurenergy .com/solar-energy-news/cerc-regulation-cross-border-trade-electricity#:~:text

=National%20Load%20Dispatch%20Centre%20shall,short%2Dterm%20open
%20access%20transactions.

Weiss, J., & Tsuchida, B. (2015). *Integrating renewable energy into the electricity grid*. Advanced Energy Economy Institute. Retrieved from https://info.aee
.net/hubfs/EPA/AEEI-Renewables-Grid-Integration-Case-Studies.pdf?t
=1440089933677.

Appendix 4

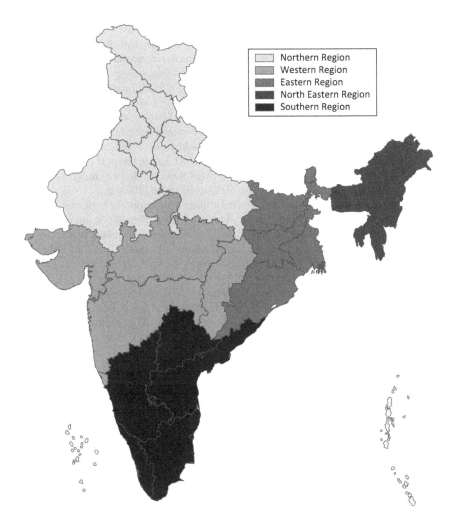

FIGURE 4.5   Five Regional Load Despatch Centres overseeing inter-state transmission.

Source: Created by authors from mapchart.net

5

ENERGY CONNECTIVITY BETWEEN INDIA AND ASEAN

Trends and Challenges

Shankaran Nambiar and Daniel Del Barrio-Alvarez

1 Introduction

India and Southeast Asia are to be at the center of the global energy demand growth in the coming years accompanying population increase, economic improvements and the universalization of electricity services. However, the available energy resources are unevenly distributed across both regions and are not always located in the countries with larger demands. For that, power connectivity is an important policy alternative for both regions to increase their energy security in a sustainable manner. So far, power connectivity has been pursued in both regions mostly separately; however, further attention is being paid to the potential of interconnections between both regions. In this process, India's North East Region (NER) and Myanmar are expected to play an important role as the gateways between South and Southeast Asia.

This chapter will be organized as follows. The next second section will discuss the need for economic cooperation between India and the Association of Southeast Asian Nations (ASEAN). It will argue that in order to sustain high levels of growth, it is necessary to improve economic cooperation trade agreements. The third section will be concerned with establishing the rationale for connectivity and regional initiatives. This will be followed by a discussion of the existing arrangements for power connectivity. The fifth section will argue that the North East Region in India and Myanmar can act as pivotal points for India and Southeast Asia to forge a link. Subsequently, an attempt is made to analyze the question of institutional development in the context of energy connectivity. Finally, a conclusion and some recommendations are offered.

DOI: 10.4324/9781003433163-5

2 Economic cooperation between India and ASEAN

South Asia in general, and India in particular, have seen rapid growth in the last decade. The same is the case with the Southeast Asian economies that have been growing rapidly despite rather soft global economic growth. Both India and ASEAN have been growing in spite of the risks and uncertainties that have faced the world, and this includes tensions in the Middle East and US–China trade tensions. The ASEAN has been expanding tremendously, integrating itself more actively into the regional and global economy. Therefore, it has seen the enlargement of regional production networks and witnessed greater connectivity.

In the last decades, South and Southeast Asia have progressed their economic cooperation and integration, but there is still space for further growth that has not been adequately explored (ADBI, 2015). Trade between South Asia and Southeast Asia has increased, and was valued at US$4 billion in 1990. It increased to US$90 billion in 2013. However, Southeast Asia's share of South Asian trade did not increase significantly, only from 6% to 10% over the same period. Investment between South and Southeast Asia also suffers from the lack of sufficient initiatives being undertaken to increase cross-regional investment. South Asia was the destination for only 9% of Southeast Asian foreign direct investment (FDI). However, South Asia regarded Southeast Asia more favorably, with 15% of total South Asian FDI being directed to ASEAN member states during the period 2009–2013.

There are many opportunities for trade and investment growth between Southeast Asia and South Asia which India's new economic diplomacy is more actively seeking (i.e. Look East and Act East policies). In order to realize these, connectivity is essential, including both "hard" and "soft" infrastructure. A number of agreements have already been put in place for economic cooperation, providing a background for power connectivity to which built upon.

The Regional Comprehensive Economic Partnership (RCEP), initiated in 2013, provides an arrangement for trade and investment liberalization between ASEAN member states and its trading partners. Among other things, this offers an important opportunity for India to further its economic integration with ASEAN. In theory, RCEP is timely and could possibly add substance to India's Look East policy and it comes at a time when ASEAN countries are planning to engage in second-generation economic reforms. In the context of ASEAN centrality, ASEAN countries wish to step up their economic growth and increase cross-regional integration so that they are a regional grouping that is a hub for investment. It is envisaged that this will make ASEAN a grouping that can compete with the likes of the North American Free Trade Agreement (NAFTA) and the EU while at the same time driving up growth for member countries. However, there are other existing bilateral agreements which will continue to facilitate energy cooperation with ASEAN.

The ASEAN–India Free Trade Agreement (AIFTA), which acts as a step toward the RCEP, has laid the foundation for cross-regional trade and investment liberalization. But there are various issues that need to be resolved. Some of the problems that need to be resolved include cross-border infrastructure links, poor trade facilitation measures, inadequate infrastructure financing, the prevalence of non-tariff barriers (NTBs) and barriers to FDI (ADBI:222). These measures have to be attended to as they will support transportation and energy connectivity.

As part of India's Act East policy, connectivity between India and Southeast Asia should be deepened in three areas. First, the overall framework of integration has to be addressed. This implies using the instrument of the AIFTA, for the moment, with the possibility of RCEP in the future. Second, Myanmar should be developed as a land bridge between India and ASEAN. The opening up of Myanmar makes it suitable as a link for ease of transportation through highways and railroads as well as the development of energy infrastructure.

Economic relations between India and ASEAN are in the early stages and there is ample scope for a deepening in cooperation and exchange. The notion of closer economic integration is well-known, and it is recognized that it brings about an expansion in the market for goods and services and increase the scope for economies of scale. Also, with more integration one can expect greater specialization, leading to more competitive industries, thereby improving regional competitiveness.

3 Rationale for power connectivity in India and Southeast Asia and the logic for inter-regional connectivity

There are multiple drivers for neighboring countries to foster connectivity between their electricity systems. Among those, differences in energy resource endowments, variations in timing of peak loads, and economies of scale in investments are commonly highlighted. Countries facing power generation shortages can increase their energy security through imports from other countries that may have generation surpluses or resources under-utilized. Similarly, time difference between countries lead to differences in the occurrences of peak loads, leading to the possibility for mutually beneficial intra-day power exchanges. Furthermore, power connectivity can help to mobilize the required investment needed for the development of generation capacity in countries with current lower energy needs. In the past, this has been particularly true for the development of hydropower projects linked to power exchange contracts.

India is forecasted to be the main source of global energy demand with an average of 2.6% annual growth through to 2030, and Southeast Asia follows with 2.5% (IEA, 2020). India can also be the hub for the import of energy into South Asia, aside from importing energy for its own use. There is, therefore, a need to match supply with demand. Since India has excess demand which it cannot meet with locally generated energy, there is a strong case for cooperation in energy and trading in energy. India is currently facing difficulties to provide a 24 × 7 supply of electricity, and these are expected to increase in the short term, a situation which has to be remedied (Parray & Tonga, 2019). This is where the argument for turning to energy resources in neighboring countries comes to the fore. Unless this option is given due attention, the shortage of energy will hinder economic growth in both regions. For an illustration of the extent of loss that will be incurred, it is worth mentioning a study by the United States Agency for International Development which estimated that planned outages in Sri Lanka and Bangladesh cost their economies a loss of about half a percentage point of GDP (USAID, 2003).

Also, it has been pointed out that regional cooperation on energy in the Greater Mekong Subregion (GMS) could reduce energy costs by nearly 20% (ADBI, 2015). This translates into a savings of US$200 billion over the period 2005–2025. The integration of energy markets in South Asia has efficiency benefits and it can yield increased revenue. It has been estimated that the potential revenue from energy trade arising from the integration of energy markets can result in revenue amounting to between US$12 billion–US$15 billion annually. The costs of ignoring energy integration and cooperation have not been calculated, but they are surely huge.

In this context, energy connectivity between India and the Greater Mekong is an interesting alternative. This connectivity would allow the synergic sharing of energy, economic, and technological resources. It would also bring further benefits by opening the door to further and strengthened collaboration between both regions; as well as be able to a source of development for the NER and Myanmar.

As India increases its generation from variable renewables, the interconnections with neighboring countries could be used both to export the surpluses and for imports when needed. India could also act as a hub for the transfer of surplus hydropower from Nepal and Bhutan to neighboring countries through the NER. This electricity trade would become very valuable for Myanmar as it requires increases in the supply of electricity, particularly during the dry months. This would give Myanmar more flexibility to plan its own expansion of electricity capacity via renewable and low carbon generation, avoiding the need to rush into carbon-intensive generation and to sign expensive emergency generation contracts.

In the medium to long term, the NER and India could also access imports from Myanmar as it is able to expand its generation capacity via solar PV, potentially hydropower (which will require time to negotiate in order to avoid damaging the ongoing peace process), and even possible imports from China and Lao PDR. India can play an important role in supporting the low carbon energy system in the region by providing always-needed financial resources, but also from its own experience in expanding low carbon generation and efficiency measures. The investments in additional generation and transmission capacity should be planned as to benefit the local population where the projects will be developed.

The increased connectivity should be accompanied by increased development opportunities. Multiple mechanisms can be put in place for that, for which, the CASA-1000 Community Support Program could serve as a model. The program supported by the World Bank and funded through the benefits of the CASA-1000 transmission line (from Kyrgyzstan and Tajikistan to Afghanistan and Pakistan) supports energy-related infrastructure investments in the communities along the transmission corridor.

The existing regional institutions such as the Greater Mekong Subregion (GMS), the South Asia Subregional Economic Cooperation (SASEC) group, and the Bay of Bengal Initiative for Multi-Sectoral Technical and Economic Cooperation (BIMSTEC) group offer opportunities for regional cooperation and a basis for expanding upon connectivity. It is this connectivity that will be the platform for energy trading, based, as it will be, on cross-border infrastructure projects. India's participation in the Bay of Bengal Initiative for Multi-Sectoral Technical and Economic Cooperation (BIMSTEC) and the South Asia Subregional Economic Cooperation (SASEC) anchor it to South Asia, while its proximity to Myanmar gives it the link to Southeast Asia over land.

Thus, India is uniquely positioned to reach out to ASEAN and provide opportunities for regional trade in energy between India, South Asia and Southeast Asia. Notwithstanding the presence of trade within these regions in energy, the above-mentioned regional institutions offer more opportunities for trade in hydropower, connection through gas pipelines and interconnection of electricity power grids. In fact, to take an example, linking the electric power grids of the GMS and SASEC will contribute to power pooling and deeper interconnection.

It is in this context that Myanmar has a substantial role to play in energy trading given its substantial reserves of hydropower capacity and natural gas. Due to its proximity to India on one hand and the rest of the ASEAN on the other, it possesses a vantage point for the location of gas pipelines. Given Myanmar's low electrification ratio (roughly 50%) (Vakulchuk et al., 2021), the immediate focus will be on increasing domestic supply, leaving the goal of cross-border trading as a medium-term objective. Indeed, for cross-border trading to take place, it is necessary that the physical and institutional infrastructure connecting Myanmar with the rest of Southeast Asia, and India should be developed.

TABLE 5.1 Electricity generation in India and selected ASEAN countries (TWh, 2016)

	Coal	Oil	Natural gas	Nuclear	Hydro	Geo-thermal	Solar PV	Wind	Biomass and waste
India	1,105	23	71	38	138	-	14	45	44
ASEAN	339	25	383	-	128	22	4.9	1.5	23
Indonesia	135	16	66	-	19	11	-	-	1.8
Malaysia	69	1.2	65	-	20	-	0.3	-	0.8
Myanmar	-	0.1	8.1	-	9.7	-	-	-	-
Philippines	43	5.7	20	-	8.1	11	1.1	1.0	0.7
Thailand	37	0.6	125	-	7.0	-	3.4	0.3	18
Viet Nam	54	1.1	46	-	64	-	-	0.2	0.1

Source: IEEJ, 2018

Myanmar can benefit substantially from opening up and becoming a bridge between South Asia and Southeast Asia (ADBI, 2015, p. 19; Florento & Corpuz, 2014). As far as energy connectivity is concerned, there are several projects that would link Myanmar with Bangladesh and India, including the Myanmar–Bangladesh–India gas pipeline project and the Tamanti hydropower project to supply electricity from Myanmar to India. But, for the proper linking of Myanmar with India, the barriers to connectivity would have to be overcome.

Myanmar has excess resources in natural gas and hydropower. The surplus hydropower that Myanmar possess can be exported to India and Bangladesh for the purpose of electricity production for export to India and Bangladesh. This requires investment in infrastructural facilities. Thus, investment in direct grid connection is required for electricity transmission on one hand and, on the other hand, specialized facilities are necessary to carry out natural gas liquefaction and regasification.

India and Myanmar have a bilateral economic partnership which provide a strong foundation upon which energy connectivity can be built. This gives a good opportunity for the NER to connect with Myanmar. FTAs are the overarching framework for institutional standards, processes and agreements which do have an impact on energy cooperation. At a less aggregative level, the resolution of regulatory barriers is important to support energy trading, which is further restricted through distorted energy pricing and the prevalence of subsidies.

Other issues that impede the flow of energy include political issues, such as security and considerations regarding the sphere of influence that countries choose to align themselves with. These issues go beyond technical and economic frameworks but are nonetheless present and have to be overcome. It is indeed possible to overcome these issues if ASEAN centrality is promoted with India as a partner in this process. Two other issues that have to be addressed include financing for infrastructure and energy projects and environmental assessments. The latter arises because environmental objections hamper the development of energy projects, although the need for environmental impact assessments cannot the denied.

In order to encourage energy trading to take place between South Asia and Southeast Asia is would be necessary, aside from improving financial support, it would be imperative to develop both the physical and institutional infrastructure. As a step toward this it would be essential to increase power pooling and energy interconnection between the two regions. Toward this end the electric power grids of the GMS and SASEC should be linked. The GMS is a good arrangement to achieve progress in energy and power trading. But it requires more institutional support in order to achieve the goals that are envisaged. Primarily, it means connecting the energy sectors in India and ASEAN, supported by subregional and international agencies.

Aside from connecting the energy sectors in India and ASEAN, it is also necessary that commercially viable energy projects be identified, as was done in the ASEAN Interconnection Master Plan Study (ASEAN Secretariat, 2011). In spite of the long gestation periods that would be involved, and the risks associated with the projects, there must be the political will to see the long-term advantages of inter-regional energy initiatives weighing over short-term problems.

4 Current situation of energy connectivity within India and Southeast Asia

Energy connectivity in Asia has emerged through subregional initiatives in Northeast, Southeast, South, and Central Asia. In major part, each region has followed their own development processes. Indeed, it has not been until recently that major efforts have been directed toward the promotion of inter-sub regional interconnections. In this content, Southeast and South Asia have emerged as the two regions with more ambitious goals in terms of regional power sector integration (ADBI, 2015; USAID, 2018).

The following subsections provide further details on each of these initiatives and on the rationale for the interconnection between South and Southeast Asia.

4.1 India, the South Asia Regional Economic Cooperation (SASEC) program and ASEAN

The need for energy cooperation has been present in South Asia going back as far as the late 1950s, and since that time there have been attempts at coming up with policies to deal with the energy supply. The issue is more pressing now with supply shortfalls and the high value of electrical outages since they are obviously factors that do not contribute to economic growth. These issues present an opportunity for the creation of a power system that stretches across the Asia Pacific region.

South Asian countries have established two major frameworks for regional cooperation in the energy sector. The South Asia Association for Regional Cooperation (SAARC) was formed in 1985 including Afghanistan, Bangladesh, Bhutan, India, Maldives, Nepal, Pakistan, and Sri Lanka. SAARC seeks to foster wider economic and political cooperation. Cooperation in the energy sector was kickstarted in 2000 with the establishment of the Technical Committee and a specialized Working Group on Energy in 2004. Furthermore, in 2014, the SAARC Intergovernmental Framework Agreement on Energy Cooperation was concluded.

In 2001, the South Asia Subregional Economic Cooperation Program was also set up by Bangladesh, Bhutan, India, Maldives, Myanmar, Nepal, and Sri Lanka. There are energy trading agreements within SASEC and with neighboring economies in electric power, petroleum products, and coal. The main trade in electricity is hydropower which India imports from Bhutan and, to a lesser extent, from Nepal. India has various projects with other South Asian countries, such as an India–Nepal transmission link in power and coal transfer through to Bhutan through India–Bangladesh links. There is also a Myanmar–India–Bangladesh gas pipeline, from the Shwe fields, through Rakhine, to Tripura in India, from which it would enter Brahmanbaria in Bangladesh, cross Bangladesh up to Jessore and enter West Bengal in India. As we can see, Myanmar is the focal point for many energy links that pass through or end up in India as they extend to Bangladesh or Southeast Asia.

The different countries in South Asia have their comparative advantage in different forms of energy. Bhutan and Nepal have an abundance of hydropower potential (around 100,000 MW) in the region and are able to supply to the surrounding countries. Similarly, Bangladesh has a large amount of natural gas (around 22.2 trillion cubic feet [tcf]). So, Bangladesh could be a supplier of gas to India and Pakistan. In that sense the region is not limited in terms of the resources that it is endowed with. However, there are political, fiscal, and infrastructure constraints that restrict the possibility of cross-border trading in energy.

India has the capability to be a hub for power in South Asia. It has cross-border interactions with South Asian countries such as Bangladesh, Bhutan, Myanmar and Nepal. The usual modus operandi for cross-border electricity trade is usually through government-initiated agreements rather than the integration of electricity grids. Nevertheless, this provides electricity for economic activities near the borders. Some of the cross-border electricity interconnections are as follows (Anbumozhi et al., 2019, p. 49):

- Bangladesh is currently connected to India through two 500 MW HVDC links of 400 kV transmission lines from Bahrampur (India) to Bheramara (Bangladesh).
- Connection with north-eastern India from 400 kV Tripura (India) to Comilla (Bangladesh).
- Nepal shares about 21 interconnections for electricity exchange with India (mostly through the Indian State of Bihar) through 11 kV and 33 kV distribution lines, and 132 kV and 400 kV transmission lines with a total capacity of up to 500 MW.
- An 11 kV distribution link from Manipur was established in India to export up to 3 MW power to Myanmar.

Other interconnections are being explored and they include possible proposals that are being examined by the India–Bangladesh Joint Technical Committees. Bangladesh has been active, and it signed an MOU with Bhutan and Nepal to facilitate power trading between Bangladesh, Bhutan and Nepal. Another agreement on power trading between Bangladesh, India and Bhutan is also on the table.

A masterplan for the transmission of power from hydropower projects in Arunachal Pradesh in the north-east salient region to other parts of India foresees the construction of a number of HVAC and HVDC lines. This line will be made to run through Bangladesh, with Bangladesh importing up to 2,000 MW. It will connect India and Sri Lanka through the towns of Madurai and Anuradhapura. This will be through a 400 kV HVDC submarine cable with a capacity of up to 1,000 MW. There are also plans to link India to Pakistan from Amritsar in India to Lahore in Pakistan.

With agreements between more countries in South Asia, often involving India, there will be more integration of participating grids. This is a necessary development since it is estimated that South Asia will require 43.2 GW additional cross-border capacity by 2036, which can be fulfilled by a more interconnected and market-oriented system.

The South Asia Subregional Economic Cooperation programs are relevant to India since they are the institution that is established to enable cross-regional connectivity. SASEC is useful to its members because it is able to access ASEAN member countries and establish trading agreements with them. Thus Bhutan, Bangladesh and India will be able to form links with other Southeast Asian countries through Myanmar.

The SASEC Energy Working Group initially met in 2011 to discuss how SASEC can motivate and cooperate in the energy sector. The objective was to determine the roadmap for energy cooperation and to determine which sectors should be targeted for cooperation. This initial meeting was followed by another in 2013, where the SASEC Electricity Transmission Utility Forum was established to coordinate the development of cross-border power transmission infrastructure. The forum's broad priority areas are development coordination, the evaluation and consideration of cross-border transmission plans and to share experiences and best practices.

Perhaps the most important on the agenda are regional grid interconnections and it is this problem that receives the most attention in the form of the South Asia Transmission Plan. SASEC's main objective presently is to provide institutional and technical support for the Transmission Plan. One outcome of the Transmission Plan is the Bangladesh–India interconnection project. Although interconnections among South Asian countries are important, their importance does not stop there. Actually, having developed interconnections within South Asia, the next step would be to link up with ASEAN. That is where the value of SASEC and ASEAN really come to the fore. Since ASEAN member states have considerable experience in energy interconnections, it would be beneficial for SASEC to learn from these practices and the ASEAN experience. Actually, in addition to the development of interconnections within countries, it is also important that the cross-regional aspects be considered.

4.2 Southeast Asia and the Greater Mekong Subregion (GMS)

Southeast Asia is the subregion that has been leading the regional power connectivity in the continent. Specifically, the continental side has reached levels of power connectivity and institutional developments like no other Asian region has achieved so far. This has mainly been achieved through the GMS Initiative, which includes Myanmar, Thailand, Lao PDR, Cambodia, Viet Nam, and the Yunnan Province and Guangxi in China. ASEAN has been very active in the promotion of power connectivity between all its members as part of its energy security and transition goals. For that, the Heads of ASEAN Power Utilities/Authorities (HAPUA) is the focal organization, with different working groups in generation and renewable energy (HAPUA working group 1); transmission and ASEAN power grid (HAPUA working group 2); distribution and power reliability and quality (HAPUA working group 3); policy studies and commercial development (HAPUA working group 4); and human resources (HAPUA working group 5). In addition to all these, currently there are several projects exploring the potential of interconnections from Australia to ASEAN countries, specifically to Indonesia and Singapore.

The Southeast Asian energy market is fast-changing since the demand for energy consumption is increasing and there is an increasing tendency to find substitutes for fossil fuels. The growth rates of countries like the Philippines, Indonesia and Viet Nam are increasing and that will require more energy to sustain these growth rates. Forecasts indicate that ASEAN's dependency on oil will increase from 44% in 2011 to 75% in 2035. With the exception of Brunei and Indonesia, the other ASEAN countries will be oil importers in the years to come. Therefore, attempts will be made to shift away from oil in the interests of energy security. Although presently the price of oil is low, this cannot be taken for granted given the volatile situation in the Middle East and the external forces affecting the oil market.

A mixture of policy responses can be anticipated in Southeast Asia. These include moving away from dependence on oil and a shift to renewable energy and environment-friendly sources of energy since ASEAN countries will want to reduce CO_2 emission levels. Two processes can be anticipated in dealing with this situation. First, ASEAN will visit the issue of energy cooperation and connectivity more seriously. Second, ASEAN will plan for energy generation on the basis of non-fossil fuel resources. ASEAN is well-suited to low-carbon energy resources such as geothermal resources, solar and biomass energy. More cost-effective hydropower is another route that some ASEAN countries can take.

ASEAN has the necessary framework to evolve a robust energy connectivity program because of the regional integration architecture that has been proposed in the ASEAN Economic Community (AEC) in 2015 and further put forward in the AEC Blueprint 2025. The AEC in 2015 had already mentioned the need for an integrated region that addresses growth challenges as well as energy security. The AEC Blueprint 2025 consists of the following five pillars (ASEAN Secretariat, 2015):

- A highly-integrated and cohesive economy.
- A competitive, innovative, and dynamic ASEAN.
- Enhanced connectivity and sectoral cooperation.
- A resilient, inclusive, people-oriented, and people-centered ASEAN.
- A global ASEAN.

Connectivity is explicitly mentioned as one of the goals that ASEAN aspires to achieve, and it includes energy connectivity since, without it, it would not be possible to achieve "a competitive, innovative, and dynamic ASEAN", nor would it be possible to achieve ASEAN centrality such that the ASEAN can fully integrate into the global economy. The AEC Blueprint 2025 explicitly mentions the ASEAN Power Grid (APG) and the Trans-ASEAN Gas Pipeline as part of the agenda for regional energy connectivity. The APG is an attempt to achieve energy interconnection of all ASEAN member states. Aside from interconnections that will link all the member states, the ASEAN Power Grid seeks to extend energy connectivity to neighboring countries such as Australia and China. China is connected via the GMS power framework.

In 2017, ASEAN exchanged about 51.7 TWh with Yunnan and Guangxi provinces. This was done through cross-border transmission lines passing through Myanmar, the Lao People's Democratic Republic and Viet Nam. Australia, for its part, is examining the possibility of exporting solar electricity to Singapore using submarine cable technology. The prospect of ASEAN emerging as a hub for power connectivity develops in view of South Asia's interest in exploring the possibility of energy cooperation using the Bay of Bengal Initiative for Multi-Sectoral Technical and Economic Cooperation as an instrument to encourage energy connectivity.

The ASEAN Plan of Action for Energy Cooperation (APAEC) 2016–2025 prioritizes the following three projects, although there are others in the pipeline (ASEAN Centre for Energy, 2015):

- System A (North System), located in Cambodia, Lao PDR, Myanmar, Thailand and Viet Nam.
- System B (South System), located in Thailand, Indonesia (Sumatra, Batam), Malaysia (Peninsular), and Singapore.
- System C (East System), located in Brunei Darussalam, Malaysia (Sabah, Sarawak), Indonesia (west and north Kalimantan) and the Philippines.

ASEAN has developed the ASEAN Plan of Action for Energy Cooperation 2010–2015. The APAEC is a plan that addresses the energy issues that are related to the ASEAN Economic Community (AEC) Blueprint 2015. The AEC, through APAEC, seeks to ensure a secure and reliable energy supply for ASEAN, aside from other means, through the ASEAN Power Grid and Trans-ASEAN Gas Pipeline (TAGP). APAEC also seeks to promote cleaner coal use, energy efficiency and conservation. It also emphasizes the need to turn to renewable energy which includes biofuels and nuclear energy as alternative sources of energy to drive economic activities and industrialization (Irawan, 2017).

The ASEAN Interconnection Master Plan Study (AIMS), which was completed in two phases, first in 2003 and then in 2010, had a proposal to set up a regional transmission network that links the ASEAN power systems. This was supposed to be undertaken in three stages, first on cross-border bilateral terms, subsequently on a subregional basis and, finally, expanding to an all-ASEAN basis that reaches out to an integrated Southeast Asian system. One outcome of this process of integration would be an ASEAN power grid system. Another proposal is electricity connection between Myanmar and Thailand (Ibrahim, 2014).

Yet another initiative undertaken on a regional basis is the Trans-ASEAN Gas Pipeline which is seen as a regional gas grid that links all the existing and planned pipeline networks by linking the existing and planned gas pipelines that belong to ASEAN members. This was a component of the ASEAN Council on Petroleum (ASCOPE)-TAGP Master Plan 2000. It involves the construction of 4,500 kilometers (km) of pipeline, worth an estimated US$7billion, that is largely supposed to be beneath the sea. The gas pipeline infrastructure had grown from 815 km in 2000 to 2,300 km of cross-border gas pipelines in 2008. This project is made up of eight bilateral gas pipeline interconnection projects. These pipelines form part of the TAGP, but all are bilateral in nature.

Transboundary power trade is quite common between countries in ASEAN. In 2016 the power trade capacity reached 5.5 GW; this is about 2% of the installed generation capacity (APAEC, 2015). The ASEAN Plan of Action for Energy Cooperation 2016–2025 is a regional initiative that attempts to go beyond bilateral arrangements for the trade in power and achieve multilateral connection frameworks.

One of the early projects within this plan is the Lao PDR, Thailand, Malaysia, Singapore Power Integration Project (LTMS-PIP). LTMS-PIP uses Thailand's transmission grid and allows Malaysia to purchase up to 100 MW of electricity power from the Lao PDR. With the launch of this project it will be possible to extend the network beyond the initial countries involved, thereby creating a network for multilateral electricity trade. This will extend the APG beyond neighboring borders (ASEAN Centre for Energy, 2017).

In 2017 Lao PDR, Thailand and Malaysia signed a cross-border power and transmission agreement, with electricity trading beginning the following year. Lao PDR began electricity trading with Malaysia using Thailand as a component in the network by sharing the transmission network. Singapore, however, has not participated in the trading arrangement yet.

While Cambodia, Lao PDR and Myanmar have good hydropower resources, there is excess demand for power in Thailand, Malaysia, Singapore and Viet Nam, creating an ideal situation for trade in power. There is, therefore, potential for a vibrant energy market in ASEAN to meet intra-regional energy needs. This, of course, implies that the appropriate domestic infrastructure be built along with the necessary cross-border interconnections.

The Greater Mekong Subregion has been leading the power connectivity in the ASEAN. Indeed, since its inception in 1992, power connectivity has been at the heart of regional cooperation in continental Southeast Asia. The regional economic cooperation program was launched in the aftermath of the Indochina wars at the initiative of the Asian Development Bank (ADB).

Energy cooperation in the GMS began as part of the GMS Economic Cooperation Program in 1992. The GMS Economic Cooperation Program was proposed as a form of economic cooperation through the construction of key shared infrastructures and to foster further political cooperation. Energy has been one of the priority sectors since its beginning. Indeed, it was already during the first ministerial meetings that led to the formal creation of the GMS program, when a list of power connectivity projects was envisioned (del Barrio Alvarez et al., 2019).

In 1995 the GMS setup the Electric Power Forum. The Electric Power Forum tried to develop the GMS power market by focusing on (i) improving the institutional framework to promote power trade, and (ii) improve infrastructure and physical interconnections to facilitate cross-border power dispatch.

In 1999, the working groups and ministerial meetings prepared and approved the Policy Statement that led to the signing, in 2002, of the intergovernmental agreement on regional power trade in the GMS (IGA) by the heads of state. The purpose of this agreement was to further power trade and harmonize the development of their power systems based on the principles of cooperation, gradualism and environmental sustainability.

Several institutional arrangements were put into place to facilitate cross-border connections and create a market for energy. In this respect, a Regional Power Trade Coordination Committee (RPTCC) was established so as to create the rules for power trade within the region. A Regional Power Trade Operating Agreement (RPTOA) was soon created and the purpose was to extend from bilateral cross-border connections to multiple seller–buyer regulatory frameworks, with the ultimate aim of creating a wholly competitive regional market. The RPTOA envisioned a four-stage process for this (ADB, 2008):

Stage 1: One-way power sales under a power purchase agreement from an independent power producer in one country to a power utility in a second country, using dedicated transmission lines established.

Stage 2: Trading between two countries, initially using spare capacity in dedicated stage 1 transmission lines, and eventually using other third-country transmission facilities.

Stage 3: All countries interconnected with 230–500 kV lines will introduce centralized operations with a regional system operator that would facilitate third-party participation in trading (entities other than generators/sellers and utilities/purchasers).

Stage 4: All countries accept legal and regulatory changes to enable a free and competitive electricity market, with independent third-party participation.

Further, the Vientiane Plan of Action for GMS Development for 2008–2012 was concluded with the intention of building the background for regional power trade. The GMS Regional Investment Framework was supposed to supersede the Vientiane Plan of Action. Various projects were planned under the new GMS Strategic Framework (2012–2022) that were to replace the Vientiane Plan of Action.

The primary focus of GMS is energy cooperation and the creation of a system that will enable regional energy trading. However, GMS is looking beyond this to take into account issues such as sustainable energy development and the use of renewable energy. As we have seen earlier, concerns about the diminishing reserves of fossil fuel have led to planning for greater energy efficiency. This has been accompanied by institutional measures to improve the capabilities of the subregion in order to act as a center for regional power trade.

In November 2013, the GMS members signed the Memorandum of Understanding for the Establishment of the Regional Power Coordination Center (RPCC). This is expected to improve the institutional structure of the GMS as far as the provision of energy is concerned and to promote a regional power market. Toward that end the REPCC will attempt to harmonize power programs, system operations, and adopt appropriate regulatory frameworks that are in keeping with a regional market. The overall agreement for the concept of regional energy cooperation can be noted from the fact that the 19th GMS Ministerial Conference supported plans for regional energy cooperation projects under the GMS Strategic Framework. The framework proposes investments in Cambodia, Lao PDR and Myanmar for national grids. The national grids will contribute to a regional grid. The national grid will also improve energy access to remote areas within the respective nations. As part of the institutional framework, working groups were also set up to consider performance standards, grid codes and on regulatory issues.

One of the reasons why a subregional approach such as the GMS is necessary is because, although the GMS is well-endowed with energy resources, different countries have different endowments. Lao PDR, Myanmar, Viet Nam and the regions of the People's Republic of China (PRC) that are within the GMS have abundant hydropower. Lao PDR and Myanmar produce more hydropower than is necessary for domestic demand. As far as electric power is concerned, Lao PDR and Myanmar have surplus electricity generated that allows them to export in excess of their domestic consumption. While Myanmar, Thailand and Viet Nam are rich in natural gas deposits, Viet Nam is rich in oil reserves. This diversity of endowments and resource advantages can be taken advantage of only if a regional cooperative approach is taken.

Since its inception, GMS countries have built several bilateral power interconnections for export of electricity from Myanmar to China, from China to Viet Nam, from Lao PDR to Thailand (there are low-tension interconnections from the export of electricity from Thailand to border towns in Lao PDR that lack connection to their national grid) and Viet Nam, and from Viet Nam to Cambodia (World Bank, 2019). This has been mostly led by export of hydropower from Lao PDR to Thailand (see Figure 5.1) and secondly to Viet Nam (see Figure 5.2). Yunnan is also gradually becoming another source of power export, mostly due to an excessive hydropower

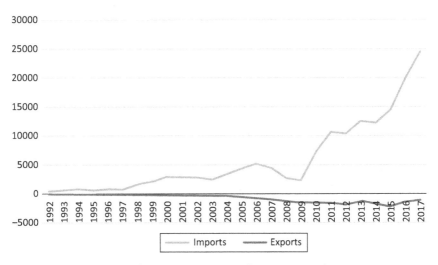

FIGURE 5.1 Thailand's electricity imports and exports (GWh).

Source: IEA (2022), Energy Statistics Data Browser, IEA, Paris https://www.iea.org/data-and-statistics/data-tools/energy-statistics-data-browser

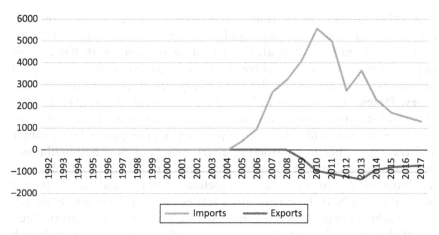

FIGURE 5.2 Viet Nam's electricity imports and exports (GWh).

Source: IEA (2022), Energy Statistics Data Browser, IEA, Paris https://www.iea.org/data-and-statistics/data-tools/energy-statistics-data-browser

capacity. China's electricity exports have been so far made only to Viet Nam, but negotiations exist to export to Myanmar. Interestingly, China has also been importing electricity from Myanmar. Cambodia is also looking to increase interconnections with neighboring counties, particularly Thailand and Lao PDR.

5 Prospects for energy connectivity between India as a member of SASEC and Southeast Asia

Opportunities for cross-regional energy trading between ASEAN and SASEC arise because energy resources are unevenly distributed in both regions. Energy resource endowments in the two regions are unevenly distributed among the regional economies, making cross-border energy projects more viable. India and Pakistan, for instance, have huge hydropower resources. However, they will still not have enough supply to meet their domestic demand as the years go by. It seems clear that this supply can come from other ASEAN countries. In that context, electricity trading between India and Myanmar can fill that gap (as would the proposed gas pipeline between Bangladesh, India and Myanmar).

A drawback of the GMS arrangement is the lack of any projects that involve cross-border projects that are able to encourage energy trade between Myanmar and South Asia. This limits the possibility of taking advantage of Myanmar as a bridge between Southeast Asia and South Asia in the matter of cross-border energy trading. This is a failing of the GMS Regional Investment Framework that otherwise has a pipeline of potential projects for 2013–2022 (GMS Secretariat, 2013).

Another problem in relation to cross-border energy trade is the difficulty with political dialogue and cooperation for energy trade. To transport gas to West Bengal, India held negotiations with Bangladesh to provide transit facilities. There was initial agreement from Bangladesh in January 2005 permitting the 895 km pipeline to pass through its territory. Despite this understanding, the government of Bangladesh kept pressing for trade concessions, leaving the conclusion of the project in abeyance. So as not to delay the project further, companies in Myanmar and India have begun considering alternative options. The possibilities that have been considered include (i) an overland route to India bypassing Bangladesh, (ii) an undersea pipeline to India, and (iii) Liquefied Natural Gas (LNG) shipments. Since settlement with Bangladesh has not been reached, the possibility of constructing a pipeline that runs through Bangladesh hangs in the balance. A pipeline that circumvents Bangladesh would be much longer (1,573 km) and costlier (US$3 billion) (World Bank 2008). Finally, to avoid the impasse, Myanmar decided to transport gas from some of the fields to the PRC (SAARC, 2010). Bangladesh, for its part, will import energy from Myanmar. The lack of political cooperation resulted in a sub-optimal solution.

There is a lot of potential to be realized by connecting the GMS and the South Asia Subregional Economic Cooperation group. That is because, as mentioned earlier, the economic potential to be gained by matching both the markets is great, and the benefits of creating an energy trading market that pools the resources of South and Southeast Asia are worth exploring. Although there is intra-regional energy trading within both South Asia and Southeast Asia, inter-regional energy trading is almost non-existent. Two aspects are missing. One is the institutional infrastructure in South Asia and the other is the volume. There is considerable volume in ASEAN and this can be tapped by India.

It is, therefore, useful to establish direct links between the GMS and SASEC so as to encourage energy trading, particularly electricity trading. However, trade in natural gas is set to emerge because natural gas can be an alternative given the rapid depletion of the world's resources in oil. Of specific interest is the abundance of natural gas in Indonesia. However, more investments in natural gas liquefaction is necessary as well as investments in regasification plants and terminals. These are essential prerequisites for trading in natural gas.

There is a lot of scope for trade in natural gas. First, due to the depletion of oil, global trade will increase by about 2% per year for the next 20 years. Second, there is a differential in natural gas resources in India and ASEAN. Most South Asian countries have very limited resources in gas. India and Bangladesh together have only 40 years of reserves; ASEAN countries, on the other hand, have about 200 years of reserves. This highlights the scope that there is for trade between India and ASEAN, as a consequence of which there is great potential for natural gas trade between South Asia and Southeast Asia. Except for Indonesia, Malaysia and Myanmar, the other economies in the two regions are either net gas importers or have no natural gas trade.

LNG trade between South Asia and ASEAN leaves much to be desired, although it is greater than that of natural gas trade. There are many constraints to the trade in LNG. These are caused because of the lack of regasification capacity. The poor LNG transport facilities and storage capacity also act as a bottleneck. Also, there are constraints in the capacity of ports to handle natural gas. Due to these constraints, South and Southeast Asian countries are not able to import sufficient LNG to meet domestic demand for gas. However, it should be noted that in order to overcome these bottlenecks, 40 new regasification plants are planned, and this includes 14 plants in Indonesia and 5 in India.

While natural gas production in Southeast Asia has more than doubled over the last two decades, the shortage of natural gas in Bangladesh, India and Pakistan has been a problem for these countries. Given the energy deficit in these countries, the lack of energy will pose a problem for their growth prospects. Bangladesh, for instance, has a natural gas deficit of 300 million standard cubic feet per day relative to a demand of 2 billion standard cubic feet per day. It has a power shortage of 1,500 MW as against a peak demand of 5,500 MW. This will make it difficult to meet demand when most required.

Indonesia and Myanmar, followed by Malaysia, are the main sources of gas in Southeast Asia, and are projected to contribute to an increase in Southeast Asian gas production from now to 2035. Total gas production in the region is forecast to grow by 30%, from 203 billion cubic meters (bcm) in 2011 to about 260 bcm in 2035 (IEA, 2013b). On the other hand, Thailand's gas production is expected to decline by 75%. The South Asian Association for Regional Cooperation (SAARC) region faces chronic electricity supply shortages. These shortages can be as severe as they are experienced in Bangladesh, where the shortfall can be as high as 28%, or be relatively mild as in Nepal (9%). The energy deficit in SAARC is because of the shortage of gas and the lack of crude oil refining capacity. Therefore, there is a shortage of petroleum products, natural gas and electricity in this region. This situation can be remedied by connecting South Asia with ASEAN. It is in this context that energy cooperation in the GMS becomes relevant to SAARC.

Energy cooperation in the GMS is important for supplying the energy needs within Southeast Asia. But given the supply constraints in South Asia, the energy that is obtained in the GMS can be applied for cross-regional energy trading. This is particularly so because of the proximity of Myanmar and the Northeast of India that link Southeast Asia with India.

Myanmar has abundant oil and gas reserves as well as hydropower potential. This abundance in energy resources places it in a suitable position to engage in cross-regional energy trade. Besides, its domestic demand is still moderate, putting it in a good position to export hydropower and natural gas. Natural gas exports, delivered through pipelines connected to gas fields off Sittwe, are already an important source of income for Myanmar, with the country supplying energy from the Shwe gas field to Thailand and China. The offshore gas fields in Yadana and Yetagun have been supplying natural gas to Thailand.

Myanmar can undertake energy trade with India and Bangladesh either through electric power transmission from hydropower projects or the transfer of natural gas. The latter can be done through pipelines or tankers. In order to take advantage of Myanmar's hydropower resources, India is developing the Tamanti multipurpose project at a site close to the India–Myanmar border. Most of the electricity generated from this project is meant for export to North East India; the Tamanti project will have substantial benefits for Myanmar, ranging from irrigation to flood control. Myanmar will benefit in other ways, the most obvious of which is through the employment opportunities that will be generated. Other opportunities include government revenue, tax income, and foreign exchange earnings. In short, this will improve Myanmar's fiscal space and improve its balance of payments. Needless to say, there will also be spinoffs through technology transfer and training, too.

Another advantage of regional integration in the GMS energy sector is the savings in total energy costs which is estimated to save about US$200 billion, or 19% of total costs. It is estimated that by extending and interlinking the interconnection of the GMS power systems it will be possible to substitute fossil fuel with hydropower. This will result in an estimated cost saving of about US$14 billion. Not only will there be a cost saving, but also the increase in carbon emissions will be greatly reduced. In this scheme of things, Myanmar has a key role to play in connecting Southeast Asia with South Asia, through India.

However, one of the limiting factors is the role played by foreign investors. In the case of electricity imports by South Asia, several of the hydropower projects are being developed as joint venture projects with foreign partners, which could aim to reserve the electricity production from such projects for their own economies, such as the PRC and Thailand. Also, the Myanmar government may give priority to expanding domestic electricity supply. Investors from Australia, Canada, the PRC, Indonesia, India, the Republic of Korea, Malaysia, Thailand and the United Kingdom are engaged in Myanmar's oil and gas sector (World Bank, 2008). Indian energy companies from both the public and private sectors have taken equity stakes for the development of gas and oil fields in Myanmar (SAARC Secretariat, 2010).

5.1 Role and meaning of power connectivity for India's North East Region

The NER has an enviable geographical location that positions it suitably within India, and also in close proximity to some of the South Asian states such as Bangladesh and Bhutan. At the same time, the NER is also in close proximity to Southeast Asia, since it is close to Myanmar. Thus, the NER can act as a connecting corridor between South and Southeast Asia, enabling energy trade along cross-border lines. The importance of the NER arises from the potential that Bhutan and Bangladesh have in increasing their energy generation potential. The latter has great potential for hydropower generation and Bangladesh for thermal power capacity. The NER itself has the capability to increase its power generation capabilities and it should, if it is to meet increasing demand for electricity within its areas and also in the surrounding region. Not only will the Northeast be a center for the generation of power, which can then be traded, but it will also be a link for cross-border energy trade.

India's Look East and Act East initiative provides a framework for the NER to transform itself and the surrounding region. However, this can be done only if there is energy sector cooperation, and the supporting infrastructure and institutional frameworks are improved and strengthened. To achieve this, several steps have to be taken, and this involves increasing the energy generation in North East India, improving infrastructure investment, connectivity and people-to-people connectivity.

This is not to deny the challenges that need to be overcome in establishing the NER as a hub for energy exchange. The primary challenge is the borders that have to be crossed in those regions for energy connectivity to be established and this brings into question a realignment in thinking on security issues, state–center relationships, technological hurdles and cross-border institutional liberalization. The other related issue concerns the natural resources, environment and settlements along those regions, particularly where energy generation is to be established and pipelines or grids constructed. Cross-border energy trading requires, as a first step, proper planning along socio-political lines in order to create the right framework for its implementation. Only then is energy integration possible.

One good reason why the capacity of the NER should be fully exploited is because with such an effort India can gradually scale down its use of coal-fired power plants, but so long as coal is used as a source of energy, there has to be a mechanism for sharing the cost of emissions in the regions. This could be based on the extent of reliance of energy from exporting countries, the logic being that emissions are increasing not only for the good of the exporting country but also for the good of the importing countries, which therefore would need to bear some portion of the costs. However, there is no doubt that there has to be a shift from fossil fuels. Bangladesh can do this by importing electricity, rather than using thermal power plants fired by coal. Nepal, Bhutan and Myanmar are known to suffer from seasonal downturns in hydropower which can be sourced from the NER if it succeeds in generating sufficient energy.

The NER is in a critical position that enables it to integrate the Bangladesh, Bhutan and Nepal (BBIN) subregion (Anbumozhi et al., 2019). This subregional arrangement is expected to work well because in the coming years Bangladesh and Nepal will become net importers of energy in view of their inability to meet their own energy needs. In that case, the NER could become the source of energy supply. But, until the NER becomes a source of energy, it has to import energy from Bhutan. Thus, one observes an increasing interaction between these countries in energy trade interactions.

There are various initiatives to further integrate the region. Many regional organizations such as SAARC, BBIN, BIMSTEC and the proposed Bangladesh–China–India–Myanmar Forum for Regional Cooperation initiative see the potential that can be realised by connecting the NER with the rest of the region. In view of this possibility several infrastructure projects have been proposed (Anbumozhi et al., 2019, p. 80). Some of the infrastructure initiatives include the following:

a. Asian Highway Link.
b. Trilateral Highways.
c. Asian Railway Network.
d. A natural gas pipeline grid.

Various models are available for the emergence of interlinkages with Southeast Asian countries. One model is through bilateral exchanges between India and Bhutan and Bangladesh and Nepal. In this model, Bhutan, which has the highest per capita consumption of energy seeks to generate 10,000 MW and export to other South Asian countries as well as Myanmar. Another model is through integration along subregional lines, particularly in the context of arrangements between the BBIN and GMS. A third model is through the creation of a regional power pool that is located in the NER–Myanmar junction. The notion of a regional energy pool has gained currency in the Nordic region and in Africa (South African Power Pool) (Anbumozhi et al., 2019, p. 3). Fourthly, it is possible to form interconnections between generators and load centers, such as between Palatana (Tripura) and Comilla (Bangladesh). This project has started exporting power to the NER and Bangladesh. Finally, energy grids could be built in the form of a "virtual energy grid" like that implemented by India between eastern and western Bhutan and the NER and connecting it with Bangladesh and Southeast Asia.

As we can see, some of the models are consistent with the Act East policy and put cooperation between India and Southeast Asia at the center of their initiatives. This is particularly so in the case of arrangements between the BBIN and GMS, which are carried out in the spirit of India–ASEAN cooperation. Creating a power pool will also connect the NER and Myanmar. The model based on virtual energy grids is yet another way to implement India–ASEAN energy linkages. Arrangements based on bilateral linkages are a less suitable way to build the energy linkages that can be strengthened by the requisite institutional platform.

The integration of Bangladesh, Bhutan, India and Nepal as a subregion is crucial as a mechanism to facilitate cross-border energy trading (CBET). In the context of BBIN, the NER is set to play a prominent role because of its geographical position and also because of its status as an energy-surplus center. The NER is calculated to have a potential of about 58,900 MW. The NER can supply as much as 40% of national needs. Since Bangladesh and Nepal as well as the surrounding states in India are likely to be energy deficient, the NER can export energy to these areas. However, for that to happen the NER infrastructure has to be upgraded and transmission networks have to be built within the NER. This will be the first step to be followed by networks linking the NER to other states within India, before then proceeding to connect with neighboring countries.

The NER can be the core of the BBIN subregion, and it can connect with Bhutan, Nepal on the South Asia side and even extend to Lao PDR by linking through Myanmar. Thus, a corridor can be created for CBET by creating power-generating hubs and transmission lines. Several pre-requisites have to be met for this strategy to materialize. First it is necessary to create the right security framework as well as develop the supporting vision from an international relations perspective. An atmosphere of mutual suspicion will damage energy cooperation. India will have to take the lead in determining the right foreign policy approach. Second, the unharnessed hydropower potential has to be tapped. Third, there should be a willingness to shift to green, renewable energy rather than depend on fossil fuels.

It should be noted that there are many regional and subregional arrangements that can make a BBIN–NER–Southeast Asia corridor work. Among the frameworks that can help in this regard are the following:

a. ASEAN–India Free Trade Area.
b. Mekong–Ganga Cooperation.
c. Bay of Bengal Initiative for Multi-Sectoral Technical and Economic Cooperation.
d. Bangladesh– China–India–Myanmar Forum for Regional Cooperation.
e. South Asian Association for Regional Cooperation.
f. Regional Comprehensive Economic Partnership.

Other initiatives such as the Central Asia–South Asia Project, the China-led growth quadrangle in the Greater Mekong Subregion, the China–Pakistan Economic Corridor, and One Belt, One Road initiatives in Asia have implications for trade and could significantly change the scope of energy trading in this region. For example, the US$60 billion + China–Pakistan Economic Corridor project, which is based on a strategy of "one corridor multiple passages", consists of 51 planned and undertaken projects; of these, 24 are energy-related, with an installed capacity of 17,608 MW. At least seven projects are now at the completion stage under its early harvest category (China–Pakistan Economic Corridor).

Initiatives taken by the Government of Bangladesh in engaging with their counterparts in Bhutan and Nepal indicate a strong possibility and acceptance on the part of India to permit the use of its grids for multiple trans-border energy flows and exchanges. In fact, this essentially bilateral framework could be a stepping stone to trilateral and multilateral frameworks for use in the BBIN subregion, and extended to other neighboring countries in Southeast Asia and beyond.

5.2 Role and meaning of power connectivity for Myanmar

Myanmar is located at the crossroads between South Asia, China and Southeast Asia. Furthermore, Myanmar participates in economic and political cooperation initiatives in both South and Southeast Asia. Myanmar is the only country that is a member simultaneously of the ASEAN, the GMS, SASEC, and BIMSTEC. In addition, Myanmar is also part of the Belt and Road Initiative.

Indeed, Myanmar is exploring the possibility of becoming such an energy bridge for inter-regional power connectivity. In the GMS, Myanmar has, so far, connected only to Yunnan Province (China) via export-oriented hydropower dams. The country had also planned several other hydropower dams to export electricity to Thailand and India. However, the difficulties faced due to local opposition has, in principle, paralyzed all these projects. The Mytsone dam in the Kachin region in Northern Myanmar has become the most visible example of the complexities faced by this type of project.

Myanmar needs to expand its power generation capacity to accommodate an increasing demand due to the economic growth and to be able to realize the goal of universal electricity access by 2030 (del Barrio-Alvarez et al., 2018b). The political reform toward the democratization of the country facilitated a re-engagement of international development partners and the development of several studies to support the energy goals. Specially, the National Electrification Plan by the World Bank, Myanmar Energy Master Plan by Asian Development Bank, and the National Electricity Master Plan by Japan International Cooperation Agency (JICA).

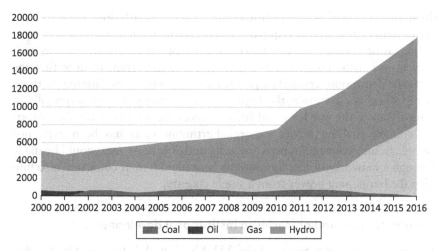

FIGURE 5.3 Electricity generation in Myanmar by source (GWh).

Source: IEA (2022), Energy Statistics Data Browser, IEA, Paris https://www.iea.org/data-and
-statistics/data-tools/energy-statistics-data-browser

The current power generation mix has traditionally been dominated by hydropower, and with an increasing role of natural gas (see Figure 5.3). This has led to a strong dependency on hydro and power shortages during dry seasons. For that, in addition to the previous studies, Myanmar is exploring other alternatives and has started to import liquefied natural gas (LNG) and launched a 1 GW solar energy tender in 2020. Indeed, Myanmar has one of the largest solar photovoltaic (PV) potentials in the region. Nevertheless, there are several elements that need to be considered in order to develop a sustainable strategy for utility-scale solar (del Barrio-Alvarez & Sugiyama, 2020).

The import of electricity from neighboring countries is one of the alternatives Myanmar is currently exploring to alleviate the current power shortages faced by Myanmar during the dry season months (del Barrio Alvarez & Sugiyama, 2018a). At least three projects for interconnections with China are being discussed. This includes a tri-national project for a China–Myanmar–Bangladesh interconnection. Talks are also being conducted for an interconnection between Lao PDR and Myanmar. All this would add to the currently existing 50 MW interconnection between Moreh (India) and Tamu (Myanmar) which has effectively promoted border electrification in Myanmar.

However, the barriers that hinder energy connectivity and trading in energy would include those that are technical, infrastructural, financial, institutional and political in nature. It should be noted that there are differences in norms and codes. This makes grid synchronization difficult. Further, convergence in grid codes to electric power is not always available, and finally there are differences in natural gas pipeline technology. An interconnection between India and Southeast Asia would necessarily run through Myanmar, which national electricity grid still does not cover the entire nation and would require reinforcements to increase its capacity.

Increasing connectivity can also confront geopolitical concerns of dependency in importing countries. Nevertheless, it is important to note that the potential disruptions created by power interconnections are different to those of pipelines. For example, the high cost of electricity storage (especially in comparison to the one for fossil fuels) reduces the incentives for the exporting country to limit power exports. Furthermore, as has been explained before, an interconnection between India and Southeast Asia would be made through a network of interconnections between different countries, also limiting the bargaining power from a single country.

6 Opportunities and obstacles to institutional development

Two significant components of the ASEAN Plan of Action are the ASEAN power grid and the ASEAN gas pipeline. What is important about both these projects is the fact that they will improve energy connectivity in ASEAN by linking up through infrastructure development with all the ASEAN member countries. Progress in meeting targets to establish the soft and hard infrastructure has been slow. A fundamental issue is that economies are more concerned with energy security at a national level and ignore possible benefits from regional energy trading (del Barrio-Alvarez & Horii, 2017). Myanmar can play an important role in taking forward ASEAN's aspirations for energy cooperation by taking a lead in promoting collective energy security. Myanmar can contribute toward the development of the objectives outlined in the ASEAN Plan of Action for Energy Cooperation 2016–2025, whose second phase, 2020–2025, was endorsed on November 2020. To address this issue, Myanmar can help promote the concept of collective energy security and stress the idea of ASEAN centrality within the field of energy by intensifying regional energy cooperation and making ASEAN less dependent on external sources for energy. Instead, ASEAN can forge links with South Asia through India as a working partner.

It is not as if the path to energy cooperation between Southeast Asia and South Asia will be smooth and efficient. Initially the link between Myanmar as the gateway for the ASEAN and India as the entry point for South Asia will have to be strengthened (ADBI, 2015, p. 71). Furthermore, it is important to not underestimate the institutional dimensions of the regional power connectivity. In this regard, existing, or future, institutional mechanisms will be required to address multiple aspects, including political, policy and regulatory, technical, economic and financial, social, environmental, and legal aspects (del Barrio-Alvarez et al., 2019).

On the technical side, institutional arrangements have to be made for the synchronization of standards and norms. This has to be undertaken for grid synchronization and grid codes as applied to power transmission and natural gas transfer. This should ensure a smooth coordination between the different rules at national and regional levels, and, in this particular case, between those in South Asia, and GMS–ASEAN. It should be not also stressed the need for information sharing mechanisms.

There are also regulatory barriers that have to be overcome in order to improve energy connectivity. There is an urgent need for the harmonization of regulations, and this is not easily accomplished because of the diversity of regulatory issues. Although Bangladesh has only one energy regulator, India and Pakistan have different regulators for each energy sector. The differences in regulatory frameworks and regulations is an obstacle to achieving progress in energy cooperation. Thus, it will not be possible to achieve a functioning regional energy market unless these differences are not resolved and a more uniform regulatory regime is gradually introduced.

There are also economic barriers to achieving energy cooperation. This is because different countries both in South Asia and in ASEAN have different economic regimes ruling the pricing of energy. Most of the countries have either a distorted energy pricing or a subsidy regime, particularly in South Asia. Thus, energy trading along market terms is not possible and if energy is sold at subsidized rates, the government will have to bear the cost of energy. On the side of consumers, it will be difficult to wean them from a regime of subsidies to a market-oriented pricing regime. Aside from the distortion to the market, caused by subsidies and price distortions, the market is also affected by the lack of access to financing. Power projects, whether they be hydropower plants or gas pipelines, require massive financing. The access to bank finance often acts as a constraint to the building of energy infrastructure. This is an issue that can be resolved through regional cooperation.

It is more difficult to find a solution to political obstacles that come in the way of energy cooperation and trade. Political barriers arise for many reasons, but most prominent is the sense that energy has consequences for the national security of a country, resulting in caution and suspicion when it comes to discussions and negotiation relating to energy issues. There are also political constraints that arise due to domestic demands on energy and domestic protests when it comes to the construction of hydropower plants and gas pipelines on the grounds of environmental destruction. The disruption to human settlements and to flora and fauna can be a problem that needs to be understood and solved, but it can result in political issues that limit regional cooperation.

The development and use of existing track II mechanisms can help to foster political trust and support to energy cooperation between both regions. In both regions, international and neutral partners have provided different fora to increase the awareness and the shared benefits that can be derived from the cooperation. In this process, two key objectives are to foster a shift in the conception from energy independence toward energy interdependence, and to promote a regional mindset in the energy planning. For that, BIMSTEC can be a platform for the coordination of support activities by relevant international partners.

There will be strong resistance to the development of domestic facilities and attempts to sell energy at market value because of vested interests. Those who have enjoyed subsidized rates will not want to transition to a system where market rates prevail. They would prefer to be able to consume energy at low rates rather than to have the energy exported. Also, many will not see the justification for large investments in the current period that will only see a return after a long time, let alone the difficulty in overcoming the credit constraints for the required investments.

As mentioned earlier, Myanmar has great potential to engage in energy trading with South Asia. Given its endowments in hydropower and natural gas, it is obvious that Myanmar can leverage these resources. It can transmit electric power from its hydropower projects and ship natural gas by tankers or via pipeline to countries like India and Bangladesh. With the intention of harnessing Myanmar's potential, India is developing the Tamanti multipurpose project, close to the India–Myanmar border. The Tamanti project is planned to develop in three stages, beginning with an installed hydropower capacity of 1,200 MW in the first stage. This will then rise to 400 MW in the second stage and finally increase to 700 MW.

Aside from developing Myanmar as a bridge, India for its part will have to develop the NER. This is a region that has been left out of the Indian development agenda for decades but is gradually receiving more attention from the government. The NER has to be the target for greater development so that it can be the hub for the region as well as the node to connect it with Myanmar and the rest of Southeast Asia (Nambiar, 2018). It is essential that there be development of infrastructure and physical connectivity to connect the cities and towns in the North East. This is a challenge given the terrain. Equally important would be to increase industrial activity and trading in the area. However, perhaps the most important is to improve stability and security in the region so as to provide a safe environment for business undertakings, for which purpose good institutions and sound governance have to be introduced.

7 Summary and recommendations

South and Southeast Asia are set to grow in the years to come; they are the new growth centers of the global economy, aside from China. This increases the need for regional cooperation and integration, which can be done through the architecture of multilateral trade agreements. More specifically, within the framework of such institutional arrangements, other initiatives are necessary. Initiatives with respect to building energy cooperation and regional initiatives are specifically necessary in order to drive growth.

In the past, energy trade between South Asia and ASEAN was limited to coal and petroleum products. Countries have pursued, with little success so far, the expansion of energy connectivity to include also natural gas pipelines and electricity trade, mostly through hydropower projects. More recently, the rapid expansion of variable renewable generation, especially solar PV but also wind, is opening opportunities for a shift toward regional energy cooperation that will support the efforts of the countries toward a transition to sustainable energy systems. It is in these areas that India and ASEAN should concentrate moving forward.

For the NER, power connectivity represents an opportunity to attract economic development as a bridge between South Asian energy exporters and importers with Southeast Asia via Myanmar (Anbumozhi et al., 2019). These interconnections will allow South Asian countries to ensure appropriate development of their renewable resources, such as hydro in Bhutan, optimize their current generation mixes, in India, and providing an alternative source of electricity via imports, in Bangladesh.

With regards to ASEAN, Myanmar is situated at a geographically strategic point given its proximity to India. Myanmar also has substantial reserves of hydropower and natural gas. This can be taken advantage of because pipelines can run from Myanmar to India or through Bangladesh. More investment is necessary in LNG liquefaction capacity and new exploration. It is also necessary to lay more natural gas pipelines, build power grids and extend hydropower facilities. For this purpose, credit availability and financing have to be extended and the credit constraint has to be relaxed. The electricity sector is another area in need of investment.

Myanmar can benefit substantially from opening up and taking advantage of its position within Southeast Asia. Myanmar could play a role by connecting through power interconnections with India and Bangladesh on one side, and Southern China, Lao PDR and Thailand on the other. In the short term, Myanmar would like to be a major importer, as well as a potential transit country for inter-regional power trade. In the long term,

Myanmar would likely become a seasonal exporter through the development of hydropower dams, and potentially a more ambitious development of utility-scale solar energy projects. Myanmar can also play a role in the trade of natural gas in the region, both from its indigenous resources, but also through the development of deep-sea ports and pipelines, such as the Kyaukpyu Deep Sea Port and the Myanmar–China gas pipeline, or the proposed Sittwe–Gaya pipeline to India. On the Eastern border, Thailand has been importing natural gas from Myanmar's Yadana, Yetagun and Zawtika gas fields. Both ASEAN and South Asia can tap into Myanmar's natural gas potential.

Aside from the infrastructure, it is also necessary to develop and liberalize trade and investment in Myanmar. This has to be in the context of GMS, SASEC and ASEAN. The norms, codes, regulatory frameworks as well as standards will have to be harmonized. BIMSTEC can provide a platform to kickstart the required initial discussions among the countries that would be at the forefront of the energy connectivity between South and Southeast Asia.

References

Anbumozhi, V., Kutani, I., & Lama, M. P. (2019). *Energising connectivity between Northeast India and its neighbours*. Jakarta: Economic Research Institute for ASEAN and East Asia.

ASEAN Centre for Energy. (2015). *ASEAN plan of action for energy cooperation*. Jakarta: ASEAN Secretariat.

ASEAN Centre for Energy. (2017). "*ASEAN Power Cooperation Report*". *ASEAN Power Cooperation Report - ASEAN Centre for Energy (aseanenergy.org)*.

ASEAN Secretariat. (2011). "*Master Plan on ASEAN Connectivity*". *Master Plan on ASEAN Connectivity.pdf*

Asia Development Bank (ADB). (2008). *Energy sector in the Greater Mekong Subregion*. Manila: ADB. Retrieved from https://www.oecd.org/countries/mongolia/42222387.pdf.

Asian Development Bank Institute (ADBI). (2015). *Connecting South Asia and SouthEast Asia*. Tokyo: Asian Development Bank Institute. https://www.adb.org/publications/connecting-south-asia-and-southeast-asia.

Del Barrio-Alvarez, D., & Hori, H. (2017). Energy security and power sector cooperation in the Greater Mekong Subregion: Past developments and near-term challenges. *Asian Journal of Public Affairs*, 9(2), e2.

Del Barrio-Alvarez, D., Numata, M., Yamaguchi, K., & Yoshikawa, H. (2018b). Universal energy access in Myanmar, challenges and opportunities. *Club Español De La Energia (Enerclub)*, 57(121), 65–72. Retrieved from https://www.enerclub.es/frontNotebookAction/Biblioteca_/Publicaciones_Enerclub/Cuadernos/CE_57_07;jsessionid=1FA526C3CF6AA3589CBFB2B0C776E998.

Del Barrio-Alvarez, D., & Sugiyama, M. (2018a). Toward a sustainable electricity policy in Myanmar: Recommendations for policy makers and development partners. Policy alternatives Research Institute - PARI policy brief. Retrieved from http://pari.utokyo.ac.jp/eng/publications/index_policy_briefs.html.

Del Barrio-Alvarez, D., & Sugiyama, M. (2020, February). A SWOT analysis of utility-scale solar in myanmar. *Energies, 13*(4), 1–17. https://doi.org/10.3390/en13040884.

Del Barrio-Alvarez, D., Yamaguchi, K., & Wang, G. (2019). The institutional dimension of power grid connectivity. In *ESCAP Electricity connectivity roadmap for Asia and the Pacific: Strategies towards interconnecting the region's grids*. https://www.unescap.org/publications/electricity-connectivity-roadmap-asia-and-pacific-strategies-towards-interconnecting.

Florento, H., & Corpuz, M. I. (2014). Myanmar: The Key Link between South Asia and Southeast Asia. ADBI Working Paper Series, No.506. Tokyo: ADBI.

Greater Mekong Subregion (GMS) Secretariat. (2013). *Regional investment framework pipeline of potential projects (2013–2022)*. Manila: Asian Development Bank. Retrieved from http://www.adb.org/publications/gms- econo mic-cooperation-program-regional-investment-framework- pipeline-projects?re f=countries/gms/publications.

Ibrahim, S. (2014, 27 March). Barriers and opportunities for electricity interconnection: The Southeast Asian experience. Presented at APERC Conference. Tokyo. Retrieved from http://aperc.ieej.or.jp/ file/2014/4/4/S2-2-2_IBRAHIM.pdf.

IEA. (2013). *Southeast Asia Energy Outlook 2013*, IEA, Paris https://www.iea.org/reports/southeast-asia-energy-outlook-2013.

IEA. (2020). *World energy Outlook*. Paris: International Energy Agency/OECD.

Irawan, B. (2017). *AEC blueprint 2025 analysis: An analysis of the ASEAN cooperation in energy*. Kuala Lumpur: CIMB ASEAN Research Institute.

Nambiar, S. (2018). India's connectivity with ASEAN: What role for Northeast India? In A. Sarma & S. Choudury (Eds.), *Mainstreaming the North east in India's look east and act east policy*. Singapore: Palgrave Macmillan.

Parray, M. T., & Tonga, R. (2019). Understanding India's power capacity: Surplus or not, and for how long? New Delhi: Brookings India.

SAARC. (2010). "SAARC Regional Energy Trade Study (SRETS)". Accessed on: https://www.sasec.asia/uploads/publications/srets_a.pdf

USAID. (2003). *"Economic Impact of Poor Power Quality on Industry". 1 (usaid. gov)*

USAID. (2018). Integrating South Asia's power grid for a sustainable and low carbon future. Retrieved from https://www.unescap.org/sites/default /files/Integrating%20South%20Asia%E2%80%99s%20Power%20Grid %20for%20a%20Sustainable%20and%20Low%20Carbon%20Future_ WEB.pdf.

Vakulchuk, R., Chan, H.Y, Rizki Kresnawan, M., Merdekawati, M., Overland, I., Fossum Sagbakken, H., Suryadi, B., Agya Utama, N., & Yurnaidi, Z. (2020). "Myanmar: How to Become an Attractive Destination for Renewable Energy Investment?" ASEAN Centre for Energy Policy Brief No 09 / 2020. DOI http://dx.doi.org/10.13140/RG.2.2.29515.00806.

World Bank. (2008). *Potential and prospects for regional energy trade in the South Asia region*. Final Report 334/08. Washington, DC: World Bank.

World Bank. (2019). Greater Mekong Subregion power market development, all business cases including the integrated GMS case. Retrieved from http://documents1.worldbank.org/curated/en/541551554971088114/pdf/Greater-Mekong-Subregion-Power-Market-Development-All-Business-Cases-including-the-Integrated-GMS-Case.pdf.

6

ENERGY CONNECTIVITY AND COOPERATION BETWEEN GULF COOPERATION COUNCIL (GCC), IRAN, AND SOUTH ASIA

Farhad Taghizadeh-Hesary, Ehsan Rasoulinezhad, and Anbumozhi Venkatachalam

1 Introduction

Energy is an essential economic commodity for all countries. Its existence can ensure reliable transportation, the welfare of households, and the capability of economic firms to produce a variety of commodities. Sorrell (2015) and Rasoulinezhad and Taghizadeh-Hesary (2020) discuss that the importance of energy for economies has seen the rate of growth of global primary energy consumption continue to grow since 1850.

Many countries in East and South Asia do not have adequate energy sources, especially fossil fuels, and are forced to import them from other countries. Generally, geographical distance from the trade partners is essential in determining import costs for these countries. Energy importers in East and South Asia try to buy energy resources from the closest energy producers to have a lower transportation cost and, therefore, lower import costs. Therefore, nearer energy producers are essential to be considered for South Asia. For instance, Iran is considered one of the suitable energy suppliers for South Asia due to its land border with Afghanistan and Pakistan and the Gulf Cooperation Council (GCC) countries. The GCC includes six energy exporting economies, i.e., Bahrain, Kuwait, Oman, Qatar, Saudi Arabia, and the UAE, established in 1981. They all have a common religion (Islam), a common language (Arabic), and are located in the Middle East. The GCC has proximity to the Indian subcontinent and has a sea border with the South Asian region. Table 6.1 represents the crude oil production of Iran and some GCC states between 1965 and 2018.

According to Table 6.1, Kuwait and Saudi Arabia were the largest crude oil producers in 1965, with a daily barrel production of 2.4 million and 2.2

DOI: 10.4324/9781003433163-6

TABLE 6.1 Total oil production in Iran and selected GCC states, 1965–2018, thousand barrels daily

	1965	1970	1980	1990	2000	2010	2014	2015	2016	2017	2018
Iran	1,908	3,848	1,479	3,270	3,850	4,421	3,714	3,853	4,586	5,024	4,715
Kuwait	2,371	3,036	1,757	964	2,244	2,556	3,097	3,061	3,141	3,001	3,049
Oman	n/a	332	285	695	955	865	943	981	1,004	971	978
Qatar	233	363	476	434	851	1,630	1,975	1,933	1,938	1,874	1,879
Saudi Arabia	2,219	3,851	10,270	7,106	9,121	9,865	11,519	11,998	12,406	11,892	12,287
UAE	282	780	1,735	1,985	2,599	2,937	3,603	3,898	4,038	3,910	3,942

Source: Authors' compilation from BP statistical review of energy (2019)

million b/d, while Saudi Arabia and Iran with crude oil production of 12.3 million and 4.7 million b/d were at the top of this list in 2018. Since 2018, however, due to the US sanctions, Iranian oil production and export have diminished drastically.[1] Other GCC states such as UAE, Kuwait, and Qatar also have significant potential in crude oil production as they produced nearly 3.9, 3.0, and 1.8 million b/d, respectively, in 2018.

Besides, these countries produce a significant volume of natural gas, indicating their potential to export these energy commodities to South Asia. Table 6.2 lists the volume of natural gas produced by Iran and selected GCC states over the period 1970–2018.

As shown in Table 6.2, among these nations, Iran has dominated the volume of gas production since 1970. The country produced 3.0 and 205.9 million tons of oil equivalent in 1970 and 2018, respectively, while Qatar and Saudi Arabia have followed Iran by producing nearly 150.9 and 96.4 million tons of oil equivalent in 2018. However, Iran consumed the major part of the produced natural gas (due to the high energy subsidies, which made the energy prices cheap and has made inefficient natural gas household consumption patterns in this country) and exported a small portion.

Table 6.3 represents the energy commodities (HS codes of 27) export volumes from GCC states to South Asian nations throughout 2001–2018.

As we can see, GCC states have experienced a significant increase in exports of crude energy commodities to South Asian countries between 2001–2018.

Generally, regarding the energy imports from the GCC states in 2018, India, Pakistan, and Bangladesh were the most important energy importers in South Asia, with an energy import volume of 24.213 billion US dollars, 2.8 billion US dollars, and 443.008 million US dollars, respectively. Furthermore, three GCC states (Saudi Arabia, UAE, and Qatar) are the major energy-exporting partners of the South Asian region in the last decade. This shows that moving to an open economy and a higher trade openness can be a useful instrument to improve the energy trade connectivity between GCC states and South Asian countries. The reasons for the better performances of these three GCC nations are their economic size, expansion of investing in energy infrastructures, the existence of sizeable proven energy reserves, and the presence of multinational energy companies in their energy industry.

Besides GCC states, Iran has experienced considerable energy trade connectivity with the South Asian region. Table 6.4 shows the energy export volumes from Iran to the region of South Asia over the period 2001–2018.

In the case of Iran–South Asia energy trade connectivity, data shown in Table 6.4 reveal that in 2001 India (with an energy import volume of 39.1 million US dollars) and Pakistan (with an energy import volume of 5.2 million US dollars) were the top energy importers from Iran among all South

TABLE 6.2 Total gas production in Iran and selected GCC states, 1965–2018, million tons of oil equivalent

	1970	1980	1990	2000	2010	2014	2015	2016	2017	2018
Iran	3.0	3.9	21.3	48.4	123.7	150.9	157.8	171.4	189.3	205.9
Bahrain	0.6	2.1	5.2	7.9	10.7	12.6	12.7	12.4	12.4	12.8
Kuwait	1.7	3.3	3.4	7.8	9.6	12.3	13.8	14.1	14.0	15.0
Oman	-	0.6	2.1	8.9	22.1	25.2	26.4	27.1	27.7	30.9
Qatar	0.9	4.2	5.6	22.2	105.9	145.8	150.5	149.4	148.2	150.9
Saudi Arabia	1.3	7.9	27.4	40.7	71.6	83.6	85.3	90.6	93.9	96.4
UAE	0.7	6.3	16.9	32.2	43.0	45.5	50.5	51.9	53.3	55.6

Source: Authors' compilation from BP statistical review of energy (2019)

TABLE 6.3 Energy export volumes from GCC states to the South Asian region, 2001–2018, thousands of US Dollar

Energy Exporter	South Asian countries	2001	2005	2010	2013	2014	2015	2016	2017	2018
GCC states	India	142,833	704,618	17,943,652	32,933,807	27,881,414	16,227,101	13,702,755	16,447,694.00	24,213,140
	Pakistan	4,936	181,543	155,153	1,600,120	1,661,208	514,866	321,238	730,492	2,849,270
	Afghanistan	300	12,830	261	32,238	37,045	5,309	6,477	21,271	38,083
	Sri Lanka	66	3,734	112,248	151,094	445,900	68,148	17,351	67,961	141,386
	Maldives	0	633	3,724	11,834	61,798	19,079	3,153	2,885	5,727
	Bhutan	0	0	0	0	0	0	0	0	0
	Bangladesh	0	9,692	22,179	101,161	46,038	39,046	36,558	52,443	443,008
	Nepal	0	2,047	0	4,492	4,790	3,720	328	486	1,654

Source: Authors' compilation from TradeMap database (www.trademap.org/Country_SelProduct_TS.aspx): Accessed Aug 2, 2020

Note: Energy commodities in this table include all commodities under HS codes 27 (2709: petroleum oils and oils obtained from bituminous minerals, crude, 2701: coal, briquettes, ovoids, and similar solid fuels manufactured from coal, 2716: electrical energy, 2711: petroleum gas and other gaseous hydrocarbons, and 2710: petroleum oils and oils obtained from bituminous minerals, not crude).

TABLE 6.4 Energy export volumes from Iran to the South Asian region, 2001–2018, thousands of US Dollar

Energy Exporter	South Asian countries	2001	2005	2010	2013	2014	2015	2016	2017	2018
Iran	India	39,098	153,454	9,426,179	8,048,391	9,545,993	4,269,284	6,749,477	9,316,990	13,226,921
	Pakistan	5,171	30,493	103,015	138,296	260,163	226,860	252,845	258,905	409,623
	Afghanistan	1,013	13,431	128,572	265,431	250,194	513,060	430,861	440,687	605,627
	Sri Lanka	370	1,500	37,079	101,507	119,762	65,792	27,461	16,333	23,775
	Maldives	0	0	521	0	0	306	30	29	119
	Bhutan	0	0	0	0	0	0	0	0	0
	Bangladesh	319	7,929	46,757	30,703	57,337	70,534	33,326	31,269	45,627
	Nepal	24	97	576	229	335	205	298	370	104

Source: Authors' compilation from TradeMap database

TABLE 6.5 BRCA for energy trade with South Asian nations

HS code	Iran	GCC states					
		Bahrain	Kuwait	Oman	Qatar	Saudi Arabia	UAE
2709	2.55	0.04	0.21	0.08	0.84	1.89	1.01
2701	0.87	0.02	0.001	0.003	0.03	0.23	0.12
2716	1.01	0.03	0.07	0.03	0.46	0.77	0.82
2711	2.32	0.07	0.05	0.54	2.04	0.18	0.09
2710	0.66	0.04	0.01	0.02	0.36	0.79	0.72

Source: Authors' calculation from raw data of Trade Map
Note 1: HS codes are: 2709: petroleum oils and oils obtained from bituminous minerals, crude, 2701: coal, briquettes, ovoids, and similar solid fuels manufactured from coal, 2716: electrical energy, 2711: petroleum gas and other gaseous hydrocarbons, and 2710: petroleum oils and oils obtained from bituminous minerals, not crude.

Asian countries and also three countries India (with an energy import volume of 13.2 billion US dollars), Afghanistan (with an energy import volume of 605.6 million US dollars) and Pakistan (with an energy import volume of 409.6 million US dollars) were the major energy importers from Iran in 2018.

Based on the data described in Table 6.5, it can be concluded that Iran and the GCC states have followed the increasing path of energy exports to the South Asian region since the early 2000s. Moreover, data show that the volume of energy exports to India (the largest economy in South Asia) from Iran is remarkably larger than from other GCC states. The greater penetration of Iran, and of UAE, Qatar, and Oman among the GCC states, in the South Asian region would be caused by the geographical position.

According to the above explanations, the importance of expanding trade ties between Iran and the GCC states as exporters of energy with South Asian countries as importers of energy is apparent. This study seeks to investigate the determinants of energy connectivity between Iran, the GCC, and the South Asian region.

Despite some earlier studies related to this topic, such as Aghevli and Sassanpour (1982), Ibrahim (1984), Chandrasekera (1986), Harami (1986), Thukral (1990), Sen (2000), Ghosh (2009), Kumar Singh (2010), Gani and Al Mawali (2013), Salehi Esfahani et al. (2013), Mahmood et al. (2014), Selvakkumaren and Limmeechokchai (2015), Kumar Shukla et al. (2017), Ahmadzai and McKinna (2018), Nowrouzi et al. (2019), Alam et al. (2019), we did not find any study focusing on analyzing and comparing energy export patterns from Iran and the GCC states to the South Asian countries. Therefore, this chapter will fill the gap in the literature.

The rest of this research is structured as follows: a brief review of the existing literature is discussed in Section 2. Section 3 represents the research methodology (BRCA index, Cosine Index, and Generalized Method of Moments–Intercountries Trade Force (GMM–ITF) gravity econometric model). Section 4 provides empirical findings. Section 5 discusses energy trade obstacles between Iran, the GCC states, and the South Asia region. The final section concludes the paper and provides some practical policy recommendations.

2 Literature review

The related literature on energy connectivity and cooperation among the GCC and between the GCC, Iran, and South Asia can be divided into three strands. The first strand contains studies concentrated on the geopolitics of energy in the GCC and Iran. The second strand focuses on earlier studies about energy trade in the South Asian region, and studies focusing on energy trade between the GCC, Iran, and South Asia are included in the third strand of literature.

Generally, the geopolitics of energy can define energy trade flows and energy cooperation among countries. The geopolitics of energy in the region of the GCC and Iran has had attention drawn to it by many scholars. Some of them tried to discuss and study the energy geopolitics of Iran. For instance, Karbassi et al. (2007) expressed that oil and gas geopolitics in Iran are complex and sensitive to various endogenous and exogenous factors. Energy depoliticization is the key element to enhance the level of energy security of Iran, which brings the energy market equilibrium in the region. Kumar Verma (2007) investigated the geopolitical energy issues of the Iran–Pakistan–India gas pipeline. The major findings proved that any energy project in this region suffers from exogenous political tensions, particularly between the US and Iran, that may negatively affect regional energy security. In another study, Bakhoda et al. (2012) argued that due to the very rich fossil energy resources in Iran, Southeast Asian nations such as India and Pakistan could improve their energy ties with Iran to ensure their energy import diversification. Bosce (2019) investigated the critical position of Iran as a potential alternative energy supplier to the EU instead of the Russian Federation. He argued that the 2014 Ukraine–Russia tension damaged the EU's trust in Russia as an energy exporter. Through the Southern Gas Corridor, Iran can take on an important role as energy supplier for the EU. In another study, Guo et al. (2019) studied the geopolitics of the energy corridor between China, Pakistan, Iran, and Turkey. They concluded that despite the current political tension in the Middle East, Iran is a unique energy transit root in the global energy market that can help to improve global energy security. Estrada et al. (2020) focused on the predicted impact of the US–Iran war on global oil price behavior. They conducted the War Oil Crisis Simulator (WOC-simulator) to check different scenarios between

1980 and 2025. They found out that any conflict between these two countries may directly lead to destabilizing global and regional energy markets.

The second strand of the existing literature focuses on energy trade patterns in South Asian countries. Mongia et al. (1994) studied the Indian need for imported energy resources to satisfy the Indian industrialization program. The major influencing factors in this issue are air pollution from fossil fuels energy resources and the price of renewable energy for Indian industries. Raj Dhungel (2008) investigated the problems and future of regional energy trade patterns in South Asia. The major results revealed that political problems are an obstacle to energy trade development in this region.

Furthermore, South Asia's cross-border energy trade has been made only between India, Bhutan, and Nepal in the last decades. Kumar Singh (2013) argued different aspects of energy security in South Asia. He concluded that India's neighbors have a vast potential in exporting energy commodities. South Asia nations should try to foster cross-border energy trade as a key solution to ensure domestic energy security. Wijayatynga et al. (2015) studied electricity trading patterns among the countries in the South Asia region. They found large-scale transmission interconnection capacity in trading hydropower in Nepal and Bhutan, which can then be transferred to India. Alam et al. (2017) tried to investigate regional power trading in South Asia. They concluded that the region of South Asia needs secured energy and power generation due to its rapid industrialization and attain sustainable development goals. Mohsin et al. (2018) analyzed the dependency level of South Asian countries on oil supply from abroad. They showed that India has a high economic resiliency against energy shocks, while Afghanistan and Bangladesh are the most oil-vulnerable countries regarding sharp energy price fluctuations. This result proves that India has a higher potential to change the foreign oil exporters to this country. Singh et al. (2018) focused on electricity trade among South Asian countries. They found out that enhancing energy sector policies in the region and increasing the number of intergovernmental energy projects may help the region to provide a reliable energy security level. Abbas et al. (2018) studied energy management issues in South Asia. They found a lack of energy management in the region, proving that South Asian countries are far behind in meeting their local energy demands, leading to a higher dependency on energy-exporting countries. Ahmadzai and McKinna (2018) investigated the opportunities and threats for Afghanistan's electrical energy. The highlight of this paper is the conclusion that Afghanistan can become a bridge between South and Central Asia to enhance the collaboration of different Asian regions in the energy sector. However, its local political tensions significantly limit Afghanistan's ability to increase Central and South Asia's energy trade and cooperation. Nag (2019) argued that South Asian countries suffer from energy deficits and energy security. He concluded that cross-border energy trade has advantages for improving energy security in the region, which is the main threat for all countries in South Asia.

Alam et al. (2019) discussed the problem of South Asia's reliance on energy imports. Similarly, they mentioned the importance of regional cooperation in ensuring energy security which can be improved by creating a mutually beneficial platform for the utilization of energy resources. Dendup and Arimura (2019) studied the opinions of Bhutan's rural population based on the 2012 Bhutan Living Standard Survey (BLSS) to consume clean cooking fuel. The results proved the importance of the level of education of information about clean energy and fossil fuels' energy resources in Bhutan's rural. Rehman et al. (2020) investigated the effects of infrastructure on the trade deficit of selected South Asian countries. The findings proved the presence of a long-run relationship between infrastructure (i.e., transport, telecommunication, energy, and financial sector) on export and trade deficit in the region. Therefore, the countries in the region should consider the improvement of infrastructures to enhance trade flows.

The third strand of the literature concentrates on earlier studies about energy trade flows between the GCC, Iran, and the South Asian region. Pandian (2005) investigated the impact of the Indo-Iran Trans-Pakistan pipeline project on the energy trade volume between India and Pakistan. The author argued that the economic relationship alone could not significantly increase energy trade volume in the South Asian region. The South Asian region needs to improve multilateralism in energy trade and energy projects with the near energy providers to that region. Verma (2007) studied the Iran–Pakistan–India gas pipeline from an energy geopolitics view. He found out that Pakistan, which refuses to establish even normal trading ties with India, craves to earn hundreds of millions of dollars in transit fees and other annual royalties from a gas pipeline that runs from Iran's South Pars fields to Barmer in western India. Nath (2014) proposed India's new inclusive trade diplomacy to increase regional trade agreements with West Asian states, particularly Iran. This finding is in line with Nathan et al. (2013), who declared that to use the energy reserves of the neighbors in western Asia, India has focused on the improvement of three gas projects, namely Iran–Pakistan–India, Turkmenistan–Afghanistan–Pakistan–India and Myanmar–Bangladesh–India. De Cordier (2016) discussed the economic relationship between Pakistan and the GCC states. The major findings revealed that the GCC might provide a high potential for energy trade flows into Pakistan and other South Asian countries. However, the main problem is that Pakistan has no common geographical border with the GCC.

Overall, it can be concluded from the existing literature that analyzing and comparing energy export patterns from Iran and the GCC states to the South Asian countries has not been considered by scholars. Hence, we try to fill in this gap in the literature by our research.

3 Research methodology

In this research, we will follow three different research approaches:

i.   Calculating the advantage or disadvantages of energy trade between Iran, the GCC, and South Asian countries based on the Ricardian trade theory using the BRCA index (Bilateral Relative Comparative Advantage Index).
ii. Calculating the complementarity of trade between Iran, the GCC, and South Asian nations to debate the possibility of energy trade potential expansion through the Cosine index.
iii. Modeling energy trade flows between Iran, the GCC, and South Asia based on the new gravity theory called Intercountries Trade Force proposed by Rasoulinezhad and Jabalameli (2019), using an econometric technique of GMM. To carry out the GMM estimation for energy trade flows between Iran, the GCC, and South Asia region, the variables obtained from the explained ITF model are employed to conduct the GMM estimation and find out the magnitudes of influencing of explanatory variables on the energy import pattern of South Asian region from Iran and the GCC states over the period 2001–2018. The GMM estimator is employed for energy trade flows from Iran and the GCC states to eight South Asian countries to estimate the coefficients of explanatory variables in the above econometric equation. The general form of GMM can be written as Equation 6.1:

$$Y_{it} = \alpha + \beta Y_{it-1} + \gamma X_{it} + \eta_{it} + \varepsilon_{it} \tag{6.1}$$

In the above equation, Y is the dependent variable (energy export flows from Iran and each GCC states to the South Asian nations) and X denotes all independent variables (gravity index, free space of trade, urbanization growth, exchange rate, geopolitical risk, geographical common border).η_{it} and ε_{it} indicate the country-specific effects and the error term, respectively.

Before carrying out the estimations, some preliminary tests should be conducted to ensure the reliability of empirical findings. The first preliminary test – Variance Inflation Factor (VIF) – is dedicated to evaluating the multicollinearity among the variables. Next, the Hausman test is employed to find out whether there is any heterogeneity in our panel. Due to the existence of different internal and external shocks in the economies of Iran, GCC states, and the South Asian nations, the next preliminary test is cross-section dependency evaluation. Finally, the 2nd generation unit root test is conducted to know the level of stationary of our series.

In addition to the mentioned pre-estimation tests, we carried out two post-estimation tests (diagnostics tests) after doing the GMM estimations. The first is the Arellano–Bond test for zero autocorrelation in the first-differenced errors, while the Sargan test is the second diagnostic test to verify the overidentifying restrictions.

4 Empirical findings

In this section, the findings of RCA, the Cosine index, and GMM estimation are represented and discussed.

4.1 The RCA index results

As we have expressed in Section 3, the RCA index in our study is calculated for five energy commodities HS code and energy exporters, i.e., Iran and GCC nations (namely Bahrain, Kuwait, Oman, Qatar, Saudi Arabia, and UAE).

According to the results shown in Table 6.5, Iran has BRCA > 1 in energy commodities of 2709 (petroleum oils and oils obtained from bituminous minerals, crude), 2716 (electrical energy), and 2711 (petroleum gas and other gaseous hydrocarbons), depicting that this country has advantages energy export flows to South Asian nations in these three commodities. Bahrain, Kuwait, and Oman are three nations of the GCC that have export disadvantages in these five energy commodities to South Asian nations. Qatar only has an export advantage in energy commodities with HS code 2711 (petroleum gas and other gaseous hydrocarbons), while Saudi Arabia and UAE have only BRCA > 1 in 2709, expressing that these two countries have a comparative advantage in exporting only petroleum oils and oils obtained from bituminous minerals, crude (HS code: 2709) to South Asian nations.

4.2 The COSINE index results

Based on the results shown in Tables 6.6–6.12, the countries of South Asia can be divided into three groups: i. the high trade potential (HTP) group, with the highest trade potential and a Cosine index range of (1–0.6); ii. the medium trade potential (MTP) group with a Cosine index range of (0.2–0.6); and iii. the low trade potential (LTP) group with a Cosine index range of (0–0.2).

The Cosine Index for Export of Energy Products from Iran to South Asia shows that the trade potential of India in importing 2709 (petroleum oils and oils obtained from bituminous minerals, crude), 2716 (electrical energy) and 2711 (petroleum gas and other gaseous hydrocarbons) is high, while in importing 2701 (coal, briquettes, ovoids, and similar solid fuels manufactured from coal) India has a medium trade potential and in 2710 (petroleum oils and oils obtained from bituminous minerals, not crude) it has low trade potential. This means that the two countries can be more active and cooperative in the field of high-potential goods codes (2709, 2716, and 2711).

The results of the Cosine index for Pakistan's importing energy commodities from Iran are different from the findings for India. Pakistan has high trade potential in imports of two energy commodities of 2716 (electrical

TABLE 6.6 Cosine Index for Export of Energy Products from Iran to South Asia, 2001–2018 Average

Countries	HS code				
	2709	2701	2716	2711	2710
India	0.81	0.31	0.62	0.65	0.11
	HTP	MTP	HTP	HTP	LTP
Pakistan	0.35	0.11	0.91	0.80	0.26
	MTP	LTP	HTP	HTP	MTP
Afghanistan	0.19	0.07	0.72	0.83	0.31
	LTP	LTP	HTP	HTP	MTP
Bangladesh	0.24	0.11	0.08	0.12	0.00
	MTP	LTP	LTP	LTP	LTP
Nepal	0.01	0.09	0.00	0.17	0.00
	LTP	LTP	LTP	LTP	LTP
Sri Lanka	0.17	0.00	0.00	0.02	0.02
	LTP	LTP	LTP	LTP	LTP
Bhutan	0.02	0.01	0.01	0.07	0.00
	LTP	LTP	LTP	LTP	LTP
Maldives	0.00	0.03	0.00	0.01	0.01
	LTP	LTP	LTP	LTP	LTP

Source: Authors' calculation from raw data of Trade Map
Note 1: HS codes are: 2709: petroleum oils and oils obtained from bituminous minerals, crude, 2701: coal, briquettes, ovoids, and similar solid fuels manufactured from coal, 2716: electrical energy, 2711: petroleum gas and other gaseous hydrocarbons, and 2710: petroleum oils and oils obtained from bituminous minerals, not crude.
Note 2: HTP, MTP, and LTP stand for high trade potential, medium trade potential, and low trade potential.

energy) and 2711 (petroleum gas and other gaseous hydrocarbons), while it has medium trade potential in 2709 (petroleum oils and oils obtained from bituminous minerals, crude) and 2710 (petroleum oils and oils obtained from bituminous minerals, not crude). Besides, based on the Cosine index results, it has low trade potential only in 2701 (coal, briquettes, ovoids, and similar solid fuels manufactured from coal). Based on the commodities with a high potential trade, it can be mentioned that the two countries have a high potential for gas pipelines (to deal trade of 2711) and electricity transmission (2716) due to their common geographical borders.

In regard to Afghanistan, the results of the Cosine index depict that this country has a high trade potential, like Pakistan, in imports of 2716 (electrical energy) and 2711 (petroleum gas and other gaseous hydrocarbons), while in the import of 2710 (petroleum oils and oils obtained from bituminous minerals, not crude), the country has a medium trade potential. In the import of two other energy commodities, namely 2709 (petroleum oils and oils obtained from bituminous minerals, crude) and 2701 (coal, briquettes, ovoids, and similar solid fuels manufactured from coal), it has a low trade potential.

TABLE 6.7 Cosine Index for Export of Energy Products from Bahrain to South Asia, 2001–2018 Average

Countries	HS code				
	2709	2701	2716	2711	2710
India	0.13	0.16	0.16	0.00	0.09
	LTP	LTP	LTP	LTP	LTP
Pakistan	0.10	0.11	0.09	0.00	0.00
	LTP	LTP	LTP	LTP	LTP
Afghanistan	0.01	0.00	0.00	0.00	0.02
	LTP	LTP	LTP	LTP	LTP
Bangladesh	0.05	0.01	0.00	0.00	0.01
	LTP	LTP	LTP	LTP	LTP
Nepal	0.09	0.10	0.09	0.01	0.00
	LTP	LTP	LTP	LTP	LTP
Sri Lanka	0.03	0.00	0.01	0.00	0.00
	LTP	LTP	LTP	LTP	LTP
Bhutan	0.04	0.01	0.00	0.00	0.00
	LTP	LTP	LTP	LTP	LTP
Maldives	0.07	0.01	0.00	0.00	0.01
	LTP	LTP	LTP	LTP	LTP

Source: Authors' calculation from raw data of Trade Map
Note 1: HS codes are: 2709: petroleum oils and oils obtained from bituminous minerals, crude, 2701: coal, briquettes, ovoids, and similar solid fuels manufactured from coal, 2716: electrical energy, 2711: petroleum gas and other gaseous hydrocarbons, and 2710: petroleum oils and oils obtained from bituminous minerals, not crude.
Note 2: HTP, MTP, and LTP stand for high trade potential, medium trade potential, and low trade potential.

The status of the Cosine index results for the other five countries in the South Asian region – Bangladesh (except for imports in 2709, i.e. petroleum oils and oils obtained from bituminous minerals, crude, which has a medium trade potential), Nepal, Sri Lanka, Bhutan, and the Maldives – is similar. All five of these countries have low trade potential in importing these five energy goods (commodities with HS codes of 2709, 2701, 2716, 2711, and 2710) from Iran, which could be due to the small size of their economy, the small share of fossil fuels in their energy consumption basket, the lack of industrialization of their economy, the long geographical distance and the lack of common geographical borders between them and Iran.

Table 6.7 shows the results of the Cosine index average for the period 2001–2008 for the export of energy commodities (HS codes 2709, 2701, 2716, 2711, and 2710) from Bahrain as an island nation in the Persian Gulf to South Asian countries. Due to various factors such as the small size of Bahrain's national economy, Bahrain's low production level and exports in

TABLE 6.8 Cosine Index for Export of Energy Products from Kuwait to South Asia, 2001–2018 Average

Countries	HS code				
	2709	2701	2716	2711	2710
India	0.21	0.08	0.00	0.03	0.11
	MTP	LTP	LTP	LTP	LTP
Pakistan	0.20	0.04	0.00	0.01	0.00
	MTP	LTP	LTP	LTP	LTP
Afghanistan	0.23	0.03	0.07	0.00	0.00
	MTP	LTP	LTP	LTP	LTP
Bangladesh	0.14	0.06	0.00	0.00	0.04
	LTP	LTP	LTP	LTP	LTP
Nepal	0.00	0.00	0.00	0.00	0.00
	LTP	LTP	LTP	LTP	LTP
Sri Lanka	0.01	0.00	0.02	0.00	0.00
	LTP	LTP	LTP	LTP	LTP
Bhutan	0.00	0.00	0.00	0.00	0.00
	LTP	LTP	LTP	LTP	LTP
Maldives	0.14	0.04	0.00	0.00	0.01
	LTP	LTP	LTP	LTP	LTP

Source: Authors' calculation from raw data of Trade Map
Note 1: HS codes are: 2709: petroleum oils and oils obtained from bituminous minerals, crude, 2701: coal, briquettes, ovoids, and similar solid fuels manufactured from coal, 2716: electrical energy, 2711: petroleum gas and other gaseous hydrocarbons, and 2710: petroleum oils and oils obtained from bituminous minerals, not crude.

the field of energy commodities and its great geographical distance from the South Asian region, no country in South Asia has medium and high trade potential in importing five groups of energy goods from this country. In the case of Bahrain, as a member of the GCC, it is better to cooperate with other major energy exporters such as Iran and Saudi Arabia due to the inability to compete in the South Asian energy market. However, due to the political challenges between Bahrain's government and Iran (Saab, 2017), it is not possible to cooperate with this country in the energy market. Another solution to Bahrain's influence in the South Asian energy market is to cooperate in energy projects in the region. For example, the National Gas Company of Bahrain (Banagas) can enhance its presence in the region by participating and investing in energy projects in countries such as Pakistan and India.

The calculation of the Cosine index for the case of Kuwait indicates that three South Asian countries, India, Pakistan, and Afghanistan, with a Cosine index of 0.21, 0.20, and 0.23, have medium trade potential in imports of 2709 (petroleum oils and oils obtained from bituminous minerals, crude) from Kuwait. However, these three countries from the South Asian region

TABLE 6.9 Cosine Index for Export of Energy Products from Oman to South Asia, 2001–2018 Average

Countries	HS code				
	2709	2701	2716	2711	2710
India	0.61	0.09	0.00	0.09	0.24
	HTP	LTP	LTP	LTP	MTP
Pakistan	0.54	0.02	0.00	0.02	0.23
	MTP	LTP	LTP	LTP	MTP
Afghanistan	0.13	0.01	0.00	0.01	0.14
	LTP	LTP	LTP	LTP	LTP
Bangladesh	0.06	0.00	0.00	0.00	0.06
	LTP	LTP	LTP	LTP	LTP
Nepal	0.04	0.00	0.00	0.02	0.04
	LTP	LTP	LTP	LTP	LTP
Sri Lanka	0.09	0.00	0.00	0.00	0.01
	LTP	LTP	LTP	LTP	LTP
Bhutan	0.07	0.00	0.00	0.01	0.12
	LTP	LTP	LTP	LTP	LTP
Maldives	0.00	0.07	0.00	0.04	0.02
	LTP	LTP	LTP	LTP	LTP

Source: Authors' calculation from raw data of Trade Map
Note 1: HS codes are: 2709: petroleum oils and oils obtained from bituminous minerals, crude, 2701: coal, briquettes, ovoids, and similar solid fuels manufactured from coal, 2716: electrical energy, 2711: petroleum gas and other gaseous hydrocarbons, and 2710: petroleum oils and oils obtained from bituminous minerals, not crude.
Note 2: HTP, MTP, and LTP stand for high trade potential, medium trade potential, and low trade potential.

have low trade potential in the other four energy commodities (2701, 2716, 2711, and 2710). There has also been low trade potential for other South Asian countries (Bangladesh, Nepal, Sri Lanka, Bhutan, and the Maldives) in importing all five energy goods (HS codes 2709, 2701, 2716, 2711, and 2710). Kuwait has vast energy resources of crude oil, which accounts for nearly 90% of export revenues. However, this country relies heavily on oil products and natural gas for electricity generation lowering its export potential in recent years (Kuwait Energy Outlook, 2019).

According to the Cosine index findings, represented in Table 6.9, India is the only country in the South Asian region with high trade potential in importing energy commodities of 2709 (petroleum oils and oils obtained from bituminous minerals, crude) from Oman. Furthermore, India captures a medium trade potential in importing 2710 (petroleum oils and oils obtained from bituminous minerals, not crude) from Oman. In the other three energy commodities (2701, 2716, and 2711), there is low trade potential in importing by India from Oman. In regards to Pakistan, it has a high trade potential in imports of 2709 (petroleum oils and oils obtained from bituminous minerals, crude) and 2710 (petroleum oil and oil obtained from bituminous minerals, not crude), while similar to India, it has a low trade potential in imports of 2701 (coal, briquettes, ovoids, and similar solid fuels manufactured from coal), 2716 (electrical energy), and 2711 (petroleum gas and other gaseous hydrocarbons) from Oman. The rest of the South Asian region (i.e., Afghanistan, Bangladesh, Nepal, Sri Lanka, Bhutan, and Maldives) records a low trade potential in imports of energy commodities from Oman.

Oman, like Bahrain and Kuwait, is classified as a small state of the GCC. But the main reason for its success in the trade of 2709 and 2710 is the nearer geographical location of Oman to the South Asian region when compared with the other GCC states. This feature helps Oman transport crude oil and petroleum products to South Asian countries, especially India and Pakistan, at a lower cost of transportation and less time than other small GCC states such as Kuwait and Bahrain.

Table 6.10 represents the 2001–2018 average calculation of the Cosine index for the export of energy commodities from Qatar to South Asia.

TABLE 6.10 Cosine Index for Export of Energy Products from Qatar to South Asia, 2001–2018 Average

Countries	HS code				
	2709	2701	2716	2711	2710
India	0.61	0.03	0.37	0.72	0.43
	HTP	LTP	LTP	HTP	MTP
Pakistan	0.56	0.01	0.48	0.72	0.52
	MTP	LTP	MTP	HTP	MTP
Afghanistan	0.49	0.00	0.61	0.39	0.15
	MTP	LTP	HTP	MTP	LTP
Bangladesh	0.16	0.00	0.19	0.14	0.32
	LTP	LTP	LTP	LTP	MTP
Nepal	0.23	0.00	0.12	0.08	0.14
	MTP	LTP	LTP	LTP	LTP
Sri Lanka	0.17	0.02	0.19	0.21	0.09
	LTP	LTP	LTP	MTP	LTP
Bhutan	0.13	0.05	0.02	0.18	0.12
	LTP	LTP	LTP	LTP	LTP
Maldives	0.03	0.00	0.06	0.05	0.04
	LTP	LTP	LTP	LTP	LTP

Source: Authors' calculation from raw data of Trade Map
Note 1: HS codes are as: 2709: petroleum oils and oils obtained from bituminous minerals, crude, 2701: coal, briquettes, ovoids, and similar solid fuels manufactured from coal, 2716: electrical energy, 2711: petroleum gas and other gaseous hydrocarbons, and 2710: petroleum oils and oils obtained from bituminous minerals, not crude.
Note 2: HTP, MTP, and LTP stand for high trade potential, medium trade potential, and low trade potential.

Qatar is one of the major exporters of oil and gas among the GCC states. According to Table 6.10, India has a high trade potential in importing two energy commodities of 2709 (petroleum oils and oils obtained from bituminous minerals, crude) and 2711 (petroleum gas and other gaseous hydrocarbons). It also has a medium trade potential in imports of 2710 (petroleum oils and oils obtained from bituminous minerals, not crude) and a low trade potential in imports of 2701 (coal, briquettes, ovoids, and similar solid fuels manufactured from coal) and 2716 (electrical energy). The findings of the Cosine index in the import of Pakistan's energy products from Qatar indicate that Pakistan has a high trade potential in purchasing and importing 2711 (petroleum gas and other gaseous hydrocarbons), while in importing three goods of 2709 (petroleum oils and oils obtained from bituminous minerals, crude), 2710 (petroleum oils and oils obtained from bituminous minerals, not crude) and 2716 (electrical energy), it has a medium trade potential and in importing 2701 (coal, briquettes, ovoids, and similar solid fuels manufactured from coal), it has low trade potential. In the case of Afghanistan, the results show that the country had a high trade potential in importing 2716 (electrical energy), and a medium trade potential of 2709 (petroleum oils, and oils obtained from bituminous minerals, crude) and 2711 (petroleum gas and other gaseous hydrocarbons) in imports. Afghanistan also has low trade potential in imports of two goods with HS codes 2701 (coal, briquettes, ovoids, and similar solid fuels manufactured from coal) and 2710 (petroleum oils and oils obtained from bituminous minerals, not crude). The other five South Asian countries (Bangladesh, Nepal, Sri Lanka, Bhutan, and the Maldives) have taken similar trade potential, except Bangladesh in the import of 2710 (petroleum oils and oils obtained from bituminous minerals, not crude), Nepal in the import of 2709 (petroleum oils and oils obtained from bituminous minerals, crude) and Sri Lanka in the import of 2711 (petroleum gas and other gaseous hydrocarbons) which had medium trade potential. According to IENE (Institute of Energy for South-East Europe), the South Asian energy market is interesting for Qatar, especially for exporting Liquefied Petroleum Gas (LPG). Qatar's oil and LPG can deliver to the Indian west coast in nearly five days, compared with over 23 days to Japan (IENE). Therefore, strengthening the trade of energy commodities with high trade potential in the countries of the South Asian region, as well as converting trade of goods with medium to high trade potential, can be considered by Qatar's managers and policymakers. Expanding negotiations between Qatari leaders and the South Asian region's policymakers and more effective energy marketing in the region could be major to Qatar's future strategies in the South Asian region.

Table 6.11 reports the results of the Cosine index for imports of energy commodities by the South Asian nations from Saudi Arabia, leading in global crude oil production and OPEC policymaking (Ashfaq et al. 2020). India,

TABLE 6.11 Cosine Index for Export of Energy Products from Saudi Arabia to South Asia, 2001–2018 Average

Countries	HS code				
	2709	2701	2716	2711	2710
India	0.90	0.08	0.11	0.09	0.62
	HTP	LTP	LTP	LTP	HTP
Pakistan	0.73	0.03	0.11	0.05	0.43
	HTP	LTP	LTP	LTP	MTP
Afghanistan	0.43	0.00	0.05	0.00	0.42
	MTP	LTP	LTP	LTP	MTP
Bangladesh	0.32	0.01	0.01	0.05	0.21
	MTP	LTP	LTP	LTP	MTP
Nepal	0.13	0.03	0.00	0.02	0.19
	LTP	LTP	LTP	LTP	LTP
Sri Lanka	0.10	0.01	0.00	0.06	0.12
	LTP	LTP	LTP	LTP	LTP
Bhutan	0.04	0.00	0.00	0.01	0.15
	LTP	LTP	LTP	LTP	LTP
Maldives	0.09	0.00	0.00	0.02	0.18
	LTP	LTP	LTP	LTP	LTP

Source: Authors' calculation from raw data of Trade Map

Note 1: HS codes are: 2709: petroleum oils and oils obtained from bituminous minerals, crude, 2701: coal, briquettes, ovoids, and similar solid fuels manufactured from coal, 2716: electrical energy, 2711: petroleum gas and other gaseous hydrocarbons, and 2710: petroleum oils and oils obtained from bituminous minerals, not crude.

Note 2: HTP, MTP, and LTP stand for high trade potential, medium trade potential and low trade potential.

as the strongest of the South Asian countries, has a high trade potential in imports of 2709 (petroleum oils and oils obtained from bituminous minerals, crude) and 2710 (petroleum oils and oils obtained from bituminous minerals, not crude) from Saudi Arabia, while due to the lack of production power of Saudi Arabia in other energy commodities, the Cosine index of India for imports of 2701 (coal, briquettes, ovoids, and similar solid fuels manufactured from coal), 2716 (electrical energy) and 2711 (petroleum gas and other gaseous hydrocarbons) show low trade potential. In regard to Pakistan, this country takes the high trade potential level in importing 2709 (petroleum oils and oils obtained from bituminous minerals, crude), but has medium trade potential and low trade potential in imports of 2710 (petroleum oils and oils obtained from bituminous minerals, not crude) and 2701 (coal, briquettes, ovoids, and similar solid fuels manufactured from coal), 2716 (electrical energy) and 2711 (petroleum gas and other gaseous hydrocarbons), respectively. Besides, Afghanistan and Bangladesh have similar trade potential in importing these five energy commodities from

Saudi Arabia. They only had two types of medium and low trade potential in importing energy goods from Saudi Arabia. They have medium trade potential in imports of 2701 coal, briquettes, ovoids, and similar solid fuels manufactured from coal) and 2710 (petroleum oils and oils obtained from bituminous minerals, not crude), and low trade potential in purchases and imports of 2701 (coal, briquettes, ovoids, and similar solid fuels manufactured from coal), 2716 (electrical energy), and 2711 (petroleum gas and other gaseous hydrocarbons). The other South Asian countries (Nepal, Sri Lanka, Bhutan, and the Maldives) have low trade potential in importing all energy commodities from Saudi Arabia. One of the main problems of Saudi Arabia is the lack of diversification in energy production. The country only focuses on the production of crude oil; for instance, gas is assumed in Saudi Arabia to be neither exported nor imported (Sarrakh et al., 2020), while India is trying to shift fossil fuels consumption to cleaner energy resources such as gas (e.g., see Purohit & Chaturvedi, 2018; D' Sa & Murthy 2004; Painuly & Parikh, 1994). Therefore, if Saudi Arabia's energy production and exports are not diversified, the trading potentials of South Asian importing countries may tend to be moderate and low in the near future.

Finally, the findings of the Cosine index of energy commodities export to South Asian countries by UAE are reported in Table 6.12. UAE is one of the main OPEC oil producers and the leading GCC states in expanding its trade potential (Khan & Alam, 2014) through the strategy of Export-Led Growth (ELG) that has been proved by Kalaitzi and Cleeve (2017). Therefore, due to the facilitation of trade and the reduction of administrative bureaucracies in the UAE, it has become the regional hub for trade (Jabado et al., 2015). As listed in Table 6.6, it is observed that the countries of the South Asian region have significant potential in the import of energy goods from this country. For example, India has high trade potential in imports of 2709 (petroleum oils and oils obtained from bituminous minerals, crude), 2716 (electrical energy), and 2711 (petroleum gas and other gaseous hydrocarbons).

In contrast, it has medium and low trade potential in imports of 2710 (petroleum oils and oils obtained from bituminous minerals, not crude) and 2701 (coal, briquettes, ovoids, and similar solid fuels manufactured from coal), respectively. In the case of Pakistan, this country has high trade potential in imports of energy commodities with HS codes 2709 (petroleum oils and oils obtained from bituminous minerals, crude), 2716 (electrical energy), 2711 (petroleum gas and other gaseous hydrocarbons), and 2710 (petroleum oils and oils obtained from bituminous minerals, not crude), while similar to India, it has low trade potential in the import of 2701 (coal, briquettes, ovoids, and similar solid fuels manufactured from coal) from UAE. In regard to Afghanistan, the calculation of the Cosine index reveals this country takes a high trade potential level only in the

TABLE 6.12 Cosine Index for Export of Energy Products from UAE to South Asia, 2001–2018 Average

Countries	HS code				
	2709	2701	2716	2711	2710
India	0.74	0.15	0.66	0.72	0.58
	HTP	LTP	HTP	HTP	MTP
Pakistan	0.68	0.16	0.70	0.62	0.63
	HTP	LTP	HTP	HTP	HTP
Afghanistan	0.52	0.14	0.47	0.63	0.52
	MTP	LTP	MTP	HTP	MTP
Bangladesh	0.31	0.15	0.38	0.21	0.39
	MTP	LTP	MTP	MTP	MTP
Nepal	0.19	0.14	0.10	0.21	0.37
	LTP	LTP	LTP	MTP	MTP
Sri Lanka	0.13	0.08	0.05	0.01	0.07
	LTP	LTP	LTP	LTP	LTP
Bhutan	0.12	0.01	0.06	0.01	0.14
	LTP	LTP	LTP	LTP	LTP
Maldives	0.04	0.05	0.00	0.00	0.00
	LTP	LTP	LTP	LTP	LTP

Source: Authors' calculation from raw data of Trade Map
Note 1: HS codes are as: 2709: petroleum oils and oils obtained from bituminous minerals, crude, 2701: coal, briquettes, ovoids, and similar solid fuels manufactured from coal, 2716: electrical energy, 2711: petroleum gas and other gaseous hydrocarbons, and 2710: petroleum oils and oils obtained from bituminous minerals, not crude.
Note 2: HTP, MTP, and LTP stand for high trade potential, medium trade potential and low trade potential.

import of 2711 (petroleum gas and other gaseous hydrocarbons). Besides, Afghanistan has a medium trade potential in imports of 2709, 2716, and 2710, while the export of 2701 from UAE to Afghanistan does not carry any significant trade potential. Bangladesh has only two types of trade potential: medium trade potential in imports of energy commodities with HS codes of 2709, 2716, 2711, and 2710, and the low trade potential for import of 2701 from UAE. The rest of the South Asian countries (i.e., Nepal, Sri Lanka, Bhutan, and the Maldives) have similar low trade potential in imports of energy commodities from UAE (with the exception of Nepal in imports of 2710 and 2711, which are classified medium trade potential).

4.3 The GMM empirical results

Before estimating our econometric IFT model with the GMM estimator, the issues of multicollinearity among the variables, heterogeneity and cross-section dependency were evaluated. The results of the VIF-test reveal low multicollinearity between the cross-sections. Besides, the findings of the Hausman test highlight the use of panel data with random effects technique for our econometric GMM estimations. Furthermore, the evaluation of the presence of cross-section dependence in the series is conducted through Cross-Section Dependence (CSD) test. The reported findings of the cross-section dependence test indicate that cross-section presents in all series in our models. Econometrically, where there is a low multicollinearity and cross-section dependence in series, checking the stationary of variables is required. To this end, the 2nd generation panel unit root test (Baltagi & Pesaran (2007) CIPS test) with the null hypothesis of all series are I(1) is employed, and the results depicted that all series in our models are I(0).

Now we can carry out the Arellano–Bond dynamic GMM estimation to analyze our models' coefficients of independent variables. The findings of GMM estimation are as follows.

Table 6.13 represents the estimation results of the Arellano–Bond GMM for the model of energy export of Iran to the South Asian region.

Firstly, the gravity index is found to be statistically significant and positive, indicating that a 1% increase in this index that comprised economic size and geographical distance leads to an increase of Iranian energy export flows to the region of South Asia by approximately 0.04%. Second, the impact of free space of trade on Iranian energy exports to the South Asia region is statistically significant and positive, supporting that any free trade space

TABLE 6.13 Arellano–Bond dynamic GMM estimation (Iran–South Asia energy trade model)

Explanatory variables	Coefficients	Significant at 1% levels
Constant	−1.339	Yes
LGI	0.043	Yes
LFST	0.21	Yes
LUR	0.84	Yes
LEX	0.11	Yes
LGEOR	−0.48	Yes
BORDER	0.01	Yes
No. of observations	144	
Periods included	18	
Cross-sections included	8	
Wald Chi2 (5)	503.15	Yes

Source: Authors' compilation

Note 1: GI = Gravity Index, FST = Free Space of Trade, UR = Urbanization growth, GEOR = geographical risk.

Note 2: (L) indicates variables in the natural logarithms.

in this region would become a good opportunity for Iran. Third, the effect of urbanization growth is found to be positive and statistically significant. Iran's energy exports to the South Asia region increase by nearly 0.8% for every 1% increase in the region's urban population. The positive relationship between urban population and import flows has been proved by Kurniawan and Managi (2018). Fourth, it is observed that the bilateral exchange rate has a positive sign, meaning that a 1% depreciation of the bilateral exchange rate in South Asian countries may decelerate the energy export of Iran to this region by approximately 0.1%. When the South Asian countries' national currencies depreciate, their import cost will increase. Fifth, the geopolitical risk has a significant and negative coefficient, meaning that by an increase in geopolitical risk, the energy export of Iran to this region may reduce by nearly 0.4%. Finally, the impact of the existence of a common geographical border is positive and statistically significant. This means that if there is a common border between Iran and countries of South Asia, the energy trade volume may increase by nearly 1 % [1% = Exp [0.01] - 1].

The results of estimating the model of Bahrain's energy export to the South Asian region, shown in Table 6.14, reveal that the Gravity index (GI) has a positive and significant coefficient. By a 1% increase in the Gravity index, the volume of energy imports of South Asia from Bahrain may increase by approximately 0.001%. Also, the variable coefficient of free space of trade is statistically significant. It has a negative sign, which means that by an increase of 1% in free space of trade in South Asia, Bahrain's energy exports volume to this region will decrease by 0.03%.

TABLE 6.14 Arellano–Bond dynamic GMM estimation (Bahrain–South Asia energy trade model)

Explanatory variables	Coefficients	Significant at 1% levels
Constant	−1.738	Yes
LGI	0.001	Yes
LFST	−0.03	Yes
LUR	0.061	Yes
LEX	0.19	Yes
LGEOR	−0.01	Yes
BORDER	0.02	No
No. of observations	144	
Periods included	18	
Cross-sections included	8	
Wald Chi2 (5)	492.04	Yes

Source: Authors' compilation

Note 1: GI = Gravity Index, FST = Free Space of Trade, UR = Urbanization growth, GEOR = geographical risk.

Note 2: (L) indicates variables in the natural logarithms.

The reason may be the lack of export agility of Bahrain and the influence of large energy exporters in the free trade space in the South Asian region. Moreover, the growth effect of urbanization in South Asian countries on the volume of their energy imports from Bahrain is positive and has a coefficient of 0.06%. Regarding the bilateral exchange rate, a positive relationship between this variable and the energy export volume of Bahrain to the South Asia region is revealed by the findings. A 1% depreciation in the national currencies of South Asian countries may cause a reduction of nearly 0.1% in Bahrain's energy export to the South Asian region. Geopolitical risk has a statistically significant and negative coefficient, while the coefficient of the common border is found to be not statistically significant for this model.

Table 6.15 represents the empirical estimation findings of Kuwait's energy export model to the South Asian region. It is proved that the GI has a positive and significant coefficient. By a 1% increase in this variable, the volume of energy imports of South Asia from Kuwait may increase by approximately 0.01%. Furthermore, the free space of trade is statistically significant. It has a small effect on South Asian imports of energy commodities from Kuwait due to the agility of other giant energy producers such as Iran and Saudi Arabia.

Moreover, the growth effect of urbanization in South Asian countries on their energy imports from Kuwait has a positive effect. In other words, by a 1% increase in the urban population in the South Asian region, the energy

TABLE 6.15 Arellano–Bond dynamic GMM estimation (Kuwait–South Asia energy trade model)

Explanatory variables	Coefficients	Significant at 1% levels
Constant	−0.392	Yes
LGI	0.011	Yes
LFST	0.00	Yes
LUR	0.233	Yes
LEX	0.101	Yes
LGEOR	−0.00	Yes
BORDER	0.06	No
No. of observations	144	
Periods included	18	
Cross-sections included	8	
Wald Chi2 (5)	563.11	Yes

Source: Authors' compilation

Note 1: GI = Gravity Index, FST = Free Space of Trade, UR = Urbanization growth, GEOR = geographical risk.

Note 2: (L) indicates variables in the natural logarithms.

imports of this region from Kuwait may increase by nearly 0.23%. According to the bilateral exchange rate coefficient, a 1% depreciation in the national currencies of the South Asian countries may lead to a decrease in energy imports from Kuwait by approximately 0.1%. Similar to the Bahrain–South Asia energy trade model's findings, geopolitical risk also has a statistically significant and negative coefficient. In contrast, the coefficient of the common border is found to be not statistically significant for this model.

The findings of Arellano–Bond dynamic GMM estimation for Oman–South Asia energy trade model (Table 6.16) depict that the GI and free space of trade have a positive and statistically significant impact, meaning that a 1% increase in these two variables may lead to energy import acceleration of South Asian region by nearly 0.09% and 0.05%, respectively. Moreover, the impacts of urbanization growth and bilateral exchange rate have been found to be positive on Oman's energy export volumes to the South Asia region. In other words, by a 1% increase in the growth of the urban population in the region of South Asia, its energy import volume may increase by nearly 0.004%. In comparison, the import volume may decrease 0.1% by a 1% depreciation in the South Asian region's national currencies. Interestingly, geographical risk has no statistically significant coefficient. It may cause by the efficient and moderate foreign policy of Oman in the last decades (Al-Maskari et al., 2019), enabling this country to face any geopolitical risk in the region proactively. The common geographical border (sea border in the case of Oman and the South Asia region) has a positive coefficient.

TABLE 6.16 Arellano–Bond dynamic GMM estimation (Oman–South Asia energy trade model)

Explanatory variables	Coefficients	Significant at 1% levels
Constant	−1.000	Yes
LGI	0.09	Yes
LFST	0.050	Yes
LUR	0.004	Yes
LEX	0.103	Yes
LGEOR	0.012	No
BORDER	0.062	Yes
No. of observations	144	
Periods included	18	
Cross-sections included	8	
Wald Chi2 (5)	562.49	Yes

Source: Authors' compilation
Note 1: GI = Gravity Index, FST = Free Space of Trade, UR = Urbanization growth, GEOR = geographical risk.
Note 2: (L) indicates variables in the natural logarithms.

TABLE 6.17 Arellano–Bond dynamic GMM estimation (Qatar–South Asia energy trade model)

Explanatory variables	Coefficients	Significant at 1% levels
Constant	−0.934	Yes
LGI	0.018	Yes
LFST	0.013	Yes
LUR	0.032	Yes
LEX	0.103	Yes
LGEOR	−0.329	Yes
BORDER	0.000	No
No. of observations	144	
Periods included	18	
Cross-sections included	8	
Wald Chi2 (5)	523.85	Yes

Source: Authors' compilation
Note 1: GI = Gravity Index, FST = Free Space of Trade, UR = Urbanization growth, GEOR = geographical risk.
Note 2: (L) indicates variables in the natural logarithms.

Table 6.17 represents the GMM results for Qatar–South Asia energy trade model. The GI has a significantly positive coefficient. In other words, by a 1% increase in this variable, Qatar's energy export to the South Asian region may go up to approximately 0.01%. The variable of free space of trade depicts a positive impact on Qatar's energy export volume to the South Asian region. Regarding urbanization growth, the result reveals the positive relationship between this variable and Qatar's energy export volume to the South Asian region. By a 1% increase in urban population growth in the South Asian region, Qatar's volume of energy export to this region may increase by about 0.03%. Moreover, geopolitical risk harms Qatar's energy export volume to the South Asian region, whereas the common geographical border coefficient is statistically insignificant.

Regarding the Saudi Arabia–South Asia energy trade model, the findings (Table 6.18) reveal that the gravity index positively influences energy import volumes of South Asian countries from Saudi Arabia. A 1% increase in this independent variable leads to energy export volume expansion of Saudi Arabia to the region by nearly 0.103%. The relationship between free space of trade and energy import volumes of South Asian countries from Saudi Arabia is positive and statistically significant. The results indicate that with a 1% increase in the presence of free space of trade in the South Asian region, Saudi Arabia may accelerate its energy export to the region by approximately 0.31%. This shows the power of Saudi Arabia to use the

TABLE 6.18 Arellano–Bond dynamic GMM estimation (Saudi Arabia–South Asia energy trade model)

Explanatory variables	Coefficients	Significant at 1% levels
Constant	−1.616	Yes
LGI	0.103	Yes
LFST	0.319	Yes
LUR	0.028	Yes
LEX	0.233	Yes
LGEOR	−0.019	Yes
BORDER	0.006	Yes
No. of observations	144	
Periods included	18	
Cross-sections included	8	
Wald Chi2 (5)	482.06	Yes

Source: Authors' compilation

Note 1: GI = Gravity Index, FST = Free Space of Trade, UR = Urbanization growth, GEOR = geographical risk.

Note 2: (L) indicates variables in the natural logarithms.

existence of space in the energy markets of the South Asia region. Moreover, the results confirm that urbanization growth and the bilateral exchange rate positively contribute to the energy export volume expansion of Saudi Arabia to the region. A 1% increase in these two variables may lead to an increase in energy export volumes of Saudi Arabia to the South Asian region by nearly 0.02% and 0.23%, respectively. Besides, the impact of geopolitical risk on energy export volume expansion of Saudi Arabia to the region is negative, and a 1% increase in geopolitical risk level is linked to a 0.01% reduction in energy export volume of Saudi Arabia to the South Asian region. The existence of a common geographical border has a positive coefficient and may lead to an increase of energy imports of South Asian countries from Saudi Arabia by approximately 6.18% [0.061 = [Exp (0.06) - 1].

Lastly, the estimated results for the case of the UAE–South Asia energy trade pattern are represented in Table 6.19. The GI has a positive and statistically significant impact on the energy export volume of UAE to the South Asian region. The estimation inferred that a 1% increase in the Gravity Index is linked with a 0.11 increase in energy export volumes of UAE into the South Asian countries. Furthermore, the coefficient of free space of trade is high and positive, indicating that UAE can penetrate and gain market share from any free space in the energy markets of South Asia. The growth in the urban population has a positive impact of 0.05% on the energy export volume of UAE to the South Asian region. In contrast, the

TABLE 6.19 Arellano–Bond dynamic GMM estimation (UAE–South Asia energy trade model)

Explanatory variables	Coefficients	Significant at 1% levels
Constant	−1.183	Yes
LGI	0.117	Yes
LFST	0.529	Yes
LUR	0.05	Yes
LEX	0.302	Yes
LGEOR	−0.001	Yes
BORDER	0.001	No
No. of observations	144	
Periods included	18	
Cross-sections included	8	
Wald Chi2 (5)	573.84	Yes

Source: Authors' compilation
Note 1: GI = Gravity Index, FST = Free Space of Trade, UR = Urbanization growth, GEOR = geographical risk.
Note 2: (L) indicates variables in the natural logarithms.

relationship between bilateral exchange rate and energy export volume of UAE to the South Asian region is positive. The positive coefficient of the bilateral exchange rate means that a 1% appreciation of the exchange rate in the South Asian region leads to an increase in the energy export volume of UAE, the region, by nearly 0.30%.

Furthermore, our empirical estimation proved that geopolitical risk has a negative and statistically significant impact on the energy export volume of UAE to the South Asian region. A 1% increase in geopolitical risk leads to a reduction of energy export volume of UAE to the South Asian region by approximately 0.001%. Finally, the common geographical border has a positive and insignificant impact on the energy export volume of UAE to the South Asian region. The main reason would be the efficient attempts of UAE to boost up advantages of trade with countries in the world. The UAE has become the leader among GCC states in attracting FDI and also the country has also tried to lower trade obstacles in the last decades.

Summarizing the signs of the coefficients in Table 6.20, it can be concluded that the Gravity index (mix of economic size and geographical distance) has positive effects on energy export volumes of Iran and all GCC states to the South Asian region. It indicates that economic growth of South Asian countries and lowering transportation costs (geographical distance is always considered a proxy for transportation cost) can play an essential positive role in boosting up the energy export flows running from Iran and

TABLE 6.20 Summarization of effects' signs of variables

Models	LGI	LFST	LUR	LEX	LGEPR	BORDER
Iran–South Asia energy trade	+	+	+	+	−	+
Bahrain–South Asia energy trade	+	−	+	+	−	NS
Kuwait–South Asia energy trade	+	+	+	+	−	NS
Oman–South Asia energy trade	+	+	+	+	NS	+
Qatar–South Asia energy trade	+	+	+	+	−	NS
Saudi Arabia–South Asia energy trade	+	+	+	+	−	+
UAE–South Asia energy trade	+	+	+	+	−	NS

Source: Authors' compilation

Note 1: GI = Gravity Index, FST = Free Space of Trade, UR = Urbanization growth, GEOR = geographical risk.

Note 2: (L) and (NS) indicate variables in the natural logarithms and are statistically insignificant, respectively.

GCC states to the region of South Asia. Furthermore, free space of trade also has a positive effect on all energy trade models of Iran and GCC countries, except for energy exports from Bahrain to South Asia, meaning that unlike Iran and the other states of the GCC, Bahrain does not have the power to use the free trade space in the South Asian energy market. Besides, summarizing the signs of coefficients of variables reveals that urbanization growth and bilateral exchange rate have a similar impact (positive) on energy export from Iran and all six GCC states to the South Asia region. Geopolitical risk is a variable negatively affecting the export of energy commodities from Iran and the GCC to the South Asian region. This means that any political tensions, wars, sanctions, etc., which reduce the national security of energy-exporting countries in the Middle East or energy-importing countries in South Asia may increase geopolitical risk and thus may reduce energy trade volume between Iran, GCC, and South Asian countries. This variable is not statistically significant in the case of energy export from Oman to the South Asian region due to this country's moderate and peaceful foreign policy. With the existence of a common geographical border, there are statistically significant and positive impacts in the case of Iran, Oman, and Saudi Arabia. At the same time, it was not significant for the energy trade of Bahrain, Kuwait, Qatar, and UAE.

5 Conclusions and policy recommendations

This research was an empirical study investigating the specifications and characteristics of Iran and GCC energy export patterns to South Asian countries. To carry out the empirical part of our research, we calculated two famous indexes, namely the BRCA and Cosine index, to evaluate the comparative advantage and trade similarity between Iran and GCC states as energy exporters and South Asian nations as energy importers. The findings of BRCA proved that Iran, Saudi Arabia, and UAE are more successful in capturing the South Asian' markets for crude oil. At the same time, Iran and Qatar have an export advantage in natural gas rather than the other energy exporters in GCC.

Besides, the results of the calculation of the Cosine index depicted that in the import of five energy goods by South Asian countries from Iran and the GCC, three levels of high, medium, and low trade potential can be defined. The results proved that the larger economies of South Asia, such as India and Pakistan, have high trade potential in importing energy commodities from larger economies, such as Iran, the United Arab Emirates, and Saudi Arabia. Moreover, GMM results concluded that the GI has positive effects on energy export volumes of Iran and all GCC states to the South Asian region, indicating that the economic growth of South Asian countries and also lowering transportation costs can play an essential positive role in boosting up the energy export flows running from Iran and GCC states to the region of South Asia.

Considering the mentioned results and conclusions, the following policy implications for improving energy trade flows between Iran, GCC, and the South Asian region are recommended:

i. The various policies to improve cross-border energy trade between Iran, GCC, and South Asia should be determined and followed by state and private sectors in their economies. Efficient policies can increase the trading connectivity among Iran, GCC, and South Asia, and in addition, they can ensure the energy security of South Asian nations.
ii. Iran and GCC states should become more agile in using free trade space in the South Asian region. In other words, as the shares of other energy exporters in the region decrease, Iran and GCC states must react quickly and efficiently to gain more market share in the South Asian region. The need for such agility in Iran and the GCC states requires a reduction in administrative bureaucracy and the promotion of energy export management.
iii. Iran and GCC states need to work together to reduce political tensions and increase regional cooperation, rather than competing for a share of the South Asian energy market. Cooperation instead of competition

will synergize their energy exports to the South Asian region. Currently, the existence of political tensions between Iran and Arab countries such as Saudi Arabia, Bahrain, and the UAE has dramatically affected their potential for energy exports to the world and the South Asian region as well.

iv. One of the essential factors influencing energy trade relations between Iran and GCC states and the South Asian region is the presence of the shadow of Western sanctions on economic activities in Iran and the region. Therefore, it is suggested that appropriate and fair trade and financial mechanisms be established between Iran, GCC states, and the South Asian region to reduce the adverse effects of these sanctions.

v. Based on the results obtained from the BRCA index, it is clear that South Asian countries do not have high trade potential in importing all energy goods. Therefore, it is suggested that Iran and GCC states focus more on exporting high-potential energy goods to the South Asian region. In this regard, improvement of production, development of export facilities, and bilateral negotiations are proposed.

vi. Since South Asian countries have long-term plans to contribute more clean energy to their national economy, it is recommended that Iran and GCC states have an efficient plan for developing and exporting cleaner fossil fuel energy resources, including natural gas, rather than exporting coal and crude oil to the region.

vii. Currently, the region of South Asia and the Middle East are experiencing various political tensions, which increase the geopolitical risk and thus reduce the volume of energy trade between Iran, GCC states, and the region of South Asia. Through further bilateral and multilateral negotiations and the selection of win–win energy diplomacy, it is proposed that countries help reduce the risk of geopolitics and thus increase the volume of energy trade between them.

viii. Some energy exporters in the GCC bloc have focused only on producing and exporting one type of energy commodity (for example, Saudi Arabia in the production and export of crude oil). Given the high commercial potential of other energy goods such as natural gas, it is recommended that energy-exporting countries (Iran and GCC countries) implement the strategy of "diversifying the export energy portfolio". Then, on the one hand, they can increase market share in the South Asia region and, on the other hand, increase their energy trade volume and the South Asian region.

ix. Iran and GCC states as energy exporters and South Asian countries as energy importers can form a "Regional Energy Union (REU)". The existence of such a union, on the one hand, reassures the importing energy countries from South Asia at more economical prices at all times (energy supply security). On the other hand, there will always be demand for energy products from Iran and GCC countries (Energy demand security). Therefore, in general, the existence of such a union will help to ensure energy security in these two regions.

Studying and evaluating energy trade patterns running from Iran and GCC states into the South Asian economies using various indexes and econometric methods shows several other methods for future studies. These include multi-criteria decision-making (MCDM) models that can be considered to evaluate energy trade patterns between Iran, GCC, and the South Asian region. We recommend that future studies use an MCDM model to analyze energy trade between Iran, GCC, and the South Asian region.

Note

1 The US withdrew from Iran's nuclear deal (JCPOA) in May 2018 which negatively affected Iran's potential for oil production.

References

Abbas, S. Z., Kousar, A., Razzaq, S., Saeed, A., Alam, M., & Mahmood, A. (2018). Energy management in South Asia. *Energy Strategy Reviews, 21*, 25–34.

Aghevli, B. B., & Sassanpour, C. (1982). Prices, output and the trade balance in Iran. *World Development, 10*(9), 791–800.

Ahmadzai, S., & McKinna, A. (2018). Afghanistan electrical energy and transboundary water systems analyses: Challenges and opportunities. *Energy Reports, 4*, 435–469.

Alam, F., Saleque, K., Alam, Q., Mustary, I., Chowdhury, H., & Jazar, R. (2019). Dependence on energy in South Asia and the need for a regional solution. *Energy Procedia, 160*, 26–33.

Alam, F., Alam, Q., Reza, S., Khurshid-ul-Alam, S. M., Saleque, K., & Chowdhury, H. (2017). Regional power trading and energy exchange platforms. *Energy Procedia, 110*, 592–596.

Al- Maskari, A., Al- Maskari, M., Alqanoobi, M., & Kunjumuhammad, S. (2019). Internal and external obstacles facing medium and large enterprises in Rusayl Industrial Estates in the Sultanate of Oman. *Journal of Global Entrepreneurship Research, 9*(1). https://doi.org/10.1186/s40497-018-0125-3.

Ashfaq, S., Maqbool, R., & Rashid, Y. (2020). Daily dataset of oil prices and stock prices for the top oil exporting and importing countries from the region of Asia. *Data in Brief, 28*, 104871. https://doi.org/10.1016/j.dib.2019.104871.

Bakhoda, H., Almassi, M., Moharamnejad, N., Moghaddasi, R., & Azkia, M. (2012). Energy production trend in Iran and its effect on sustainable development. *Renewable and Sustainable Energy Reviews, 16*(2), 1335–1339.

Baltagi, B.H. & Pesaran M.H. (2007). Heterogeneity and cross section dependence in panel data models: Theory and applications. *Journal of Applied Economics, 22*, 229–232.

Bocse, A. (2019). EU energy diplomacy: Searching for new suppliers in Azerbaijan and Iran. *Geopolitics, 24*(1), 145–173.

Chandrasekera, D. (1986). Outlook for petroleum products in Sri Lanka. *Energy, 11*(4–5), 523–533.

De Cordier, B. (2016). The interaction between Pakistan and the countries of the Gulf Cooperation Council: 'sub-imperialism by complementarity'? *Journal of Conflict Transformation & Security, 5*(1), 7–30.

Dendup, N., & Arimura, T. (2019). Information leverage: The adoption of clean cooking fuel in Bhutan. *Energy Policy, 125*, 181–195.

D' Sa, A., & Murthy, N. (2004). LPG as a cooking fuel option for India. *Energy for Sustainable Development, 8*(3), 91–106.

Estrada, M., Park, D., Tahir, M., & Khan, A. (2020). Simulations of US-Iran war and its impact on global oil price behavior. *Borsa Istanbul Review*. In Press. https://doi.org/10.1016/j.bir.2019.11.002

Gani, A., & Al Mawali, N. (2013). Oman's trade and opportunities of integration with the Asian economies. *Economic Modelling, 31*, 766–774.

Ghosh, S. (2009). Import demand of crude oil and economic growth: Evidence from India. *Energy Policy, 37*(2), 699–702.

Guo, F. F., Huang, C. F., & Wu, X. L. (2019). Strategic analysis on the construction of new energy corridor China–Pakistan–Iran–Turkey. *Energy Reports, 5*, 828–841.

Harami, K. A. (1986). Kuwait's petroleum trade with Asia. *Energy, 11*(4–5), 405–407.

Ibrahim, S. (1984). Energy in the Arab world. *Energy, 9*(3), 217–238.

IENE. Retrieved April 7, 2020, from https://www.iene.eu/qatar-boosts-market-share-in-growing-south-asia-lpg-market-p1841.html.

Jabado, R. W., Al Ghais, S. M., Hamza, W., Henderson, A. C., Spaet, J. L., Shivji, M. S., & Hanner, R. H. (2015). The trade in sharks and their products in the United Arab Emirates. *Biological Conservation, 181*, 190–198.

Kalaitzi, A., & Cleeve, E. (2017). Export-led growth in the UAE: Multivariate causality between primary exports, manufactured exports and economic growth. *Eurasian Business Review, 8*, 341–365.

Karbassi, A. R., Abduli, M. A., & Abdollahzadeh, E. M. (2007). Sustainability of energy production and use in Iran. *Energy Policy, 35*(10), 5171–5180.

Khan, S., & Alam, S. (2014). Kingdom of Saudi Arabia: A potential destination for medical tourism. *Journal of Taibah University Medical Sciences, 9*(4), 257–262.

Kumar Shukla, A., Sudhakar, K., & Baredar, P. (2017). Renewable energy resources in South Asian countries: Challenges, policy and recommendations. *Resource-Efficient Technologies, 3*(3), 342–346.

Kumar Singh, B. (2010). *India's energy security: The changing dynamics*. London: Motilal U K Books of India.

Kumar Singh, B. (2013). South Asia energy security: Challenges and opportunities. *Energy Policy, 63*, 458–468.

Kumar Verma, S. (2007). Energy geopolitics and Iran-Pakistan-India gas pipeline. *Energy Policy, 35*(6), 3280–3301.

Kurniawan, R., & Managi, S. (2018). Coal consumption, urbanization, and trade openness linkage in Indonesia. *Energy Policy, 121*, 576–583.

Kuwait Energy Outlook 2019. Retrieved April 6, 2020, from https://www.undp.org/content/dam/rbas/doc/Energy%20and%20Environment/KEO_report_English.pdf.

Mahmood, A., Javaid, N., Zafar, A., Riaz, R., Ahmed, S., & Razzaq, S. (2014). Pakistan's overall energy potential assessment, comparison of LNG, TAPI and IPI gas projects. *Renewable and Sustainable Energy Reviews, 31*, 182–193.

Mohsin, M., Zhou, P., Iqbal, N., & Shah, S. A. (2018). Assessing oil supply security of South Asia. *Energy, 155*, 438–447.

Mongia, N., Sathaye, J., & Mongia, P. (1994), Energy use and carbon implications in India focus on industry. *Energy Policy, 22*(11), 894–906.

Nag, T. (2019). Barriers to cross-border energy cooperation and implications on energy security: An Indian perspective with reference to energy trade in South Asia. *Global Business Review.* https://doi.org/10.1177/0972150919826380. Retrieved from https://journals.sagepub.com/doi/abs/10.1177/0972150919826380.

Nath, S. (2014). Strategic partnership for economic development: India's New inclusive trade diplomacy. *Procedia, 157*, 236–243.

Nathan, H., Kulkarni, S., & Ahuja, D. R. (2013). Pipeline politics—A study of India's proposed cross border gas projects. *Energy Policy, 62*, 145–156.

Nowrouzi, A., Panahi, M., Ghaffarzadeh, H., & Ataei, A. (2019). Optimizing Iran's natural gas export portfolio by presenting a conceptual framework for non-systematic risk based on portfolio theory. *Energy Strategy Reviews, 26*, 100403. https://doi.org/10.1016/j.esr.2019.100403.

Painuly, J. P., & Parikh, J. (1994). Policy analysis of oil substitution by natural gas in India: Transport and industry sectors. *Energy Policy, 21*(1), 43–52. https://doi.org/10.1016/0301-4215(93)90207-V

Pandian, S. G. (2005). Energy trade as a confidence-building measure between India and Pakistan: A study of the Indo-Iran trans-Pakistan pipeline project. *Contemporary South Asia, 14*(3), 307–320.

Purohit, P., & Chaturvedi, V. (2018). Biomass pellets for power generation in India: A techno-economic evaluation. *Environmental Science and Pollution Research, 25*(29), 29614–29632.

Raj Dhungel, K. (2008). Regional energy trade in South Asia: Problems and prospects. *South Asia Economic Journal, 9*(1), 173, 193.

Rasoulinezhad, E., & Jabalameli, F. (2019). Russia-EU gas game analysis: Evidence from a new proposed trade mode. *Environmental Science and Pollution Research, 26*(24), 24482–24488.

Rasoulinezhad, E., & Taghizadeh-Hesary, F. (2020). Analyzing energy transition patterns in Asia: Evidence of differences in incomes levels. *Frontiers in Energy Research, 8*, 162.

Rehman, F., Noman, A., & Ding, Y. (2020). Does infrastructure increase exports and reduce trade deficit? Evidence from selected South Asian countries using a new Global Infrastructure Index. *Journal of Economic Integration.* https://doi.org/10.1186/s40008-020-0183-x. Retrieved from https://journalofeconomicstructures.springeropen.com/articles/10.1186/s40008-020-0183-x#citeas.

Saab, B. (2017). Iran's long game in Bahrain. *Atlantic Council Paper.* Retrieved April 6, 2020, from https://www.atlanticcouncil.org/in-depth-research-reports/issue-brief/iran-s-long-game-in-bahrain/.

Esfahani, H. S., Mohaddes, K., & Pesaran, M. H. (2013). Oil exports and the Iranian economy. *Quarterly Review of Economics and Finance, 53*(3), 221–237.

Sarrakh, R., Renukappa, S., Suresh, S., & Mushatat, S. (2020). Impact of subsidy reform on the kingdom of Saudi Arabia's economy and carbon emissions. *Energy Strategy Reviews, 28*, 100465. https://doi.org/10.1016/j.esr.2020.100465

Selvakkumaren, S., & Limmeechokchai, B. (2015). Low carbon scenario for an energy import-dependent Asian country: The case study of Sri Lanka. *Energy Procedia, 79,* 1033–1038.

Sen, A. (2000). Natural gas imports into South Asia a study in international relations. *Energy Policy, 28*(11), 763–770.

Singh, A., Jamasb, T., Nepal, R., & Toman, M. (2018). Electricity cooperation in South Asia: Barriers to cross-border trade. *Energy Policy, 120,* 741–748.

Sorrell, S. (2015). Reducing energy demand: A review of issues, challenges and approaches. *Renewable and Sustainable Energy Reviews, 47,* 74–82.

Thukral, K. (1990). India's oil policy: To import crudes or products? *Energy Policy, 18*(4), 368–380.

Wijayanunga, P., Chattopadhyay, D., & Fernando, P. N. (2015). Cross-border power trading in South Asia: A Techno-Economic Rationale, ADB Working Paper No. 38. Retrieved April 12, 2020, from https://www.adb.org/sites/default/files/publication/173198/south-asia-wp-038.pdf.

7

ELECTRICITY SYSTEM INTEGRATION IN EUROPE AND NORTH AMERICA

Govinda Timilsina and Sunil Malla

1 Introduction

The European Union (EU) and North America are the most advanced regions for electricity market integration. The EU electricity market[1] is fully interconnected and operates under six regional synchronous (50 Hz) operating zones or grids. They are Continental Europe, Nordic, Baltic, British, Ireland, and Northern Ireland zones. These regional zones are also interconnected to each other through direct current (DC) or alternative current (AC) asynchronous links. Turkey and North Africa (Algeria, Morocco, and Tunisia) are also interconnected, through synchronous links, with the European power system. The only two EU Member States operating in isolation from the EU's interconnected power system are island countries – Cyprus and Iceland. The European power system is the largest interconnected power system in the world operating in synchronous mode. In North America, Canada and the US are fully interconnected, whereas the interconnection between Mexico and the US is limited. The integrated power systems of Canada and the US are divided into five regional grids. They are the Eastern, Western, Texas, Québec, and Alaska interconnection grids. Each grid operates in synchronous mode with a 60 Hz frequency. These grids are interconnected through AC or DC links.

The EU and North America can serve as global models of highly functional, cross-border electricity coordination. Both regions have more than a century of experience in cross-border electricity cooperation, interconnection, and trade. They exhibit examples of successful coordination of power system operations, power system development and expansion policies, and institutional/regulatory setups to govern the power systems. As a result, these systems are efficient and reliable to provide electricity services to their customers.

DOI: 10.4324/9781003433163-7

More recently, cross-border electricity interconnection and regional electricity trade have become critical avenues for contributing to the global efforts for climate change mitigation. From the climate change perspective, the importance of power system interconnections and trade is driven by two facts. First, the power sector is the primary source of greenhouse gas (GHG) emissions due to the widespread use of fossil fuels for power generation. Second, cross-border electricity interconnection and regional trade facilitate the substitution of fossil fuels in a region with cleaner sources of electricity generation, such as hydro, solar, and wind in the neighboring regions. Climate change mitigation has become the main driver of cross-border electricity interconnection and regional electricity trade in both Europe and North America, besides the other drivers mentioned above. The EU and North America (the US, Canada, and Mexico) have set targets of reducing their GHG emissions under the Paris Climate Agreement and planning to meet these targets through actions under their Nationally Determined Contributions (NDC).[2]

The cross-border electricity trade in North America provides an opportunity to help reduce the region's GHG emissions through the exploitation of Canada's hydropower resources. Canada has more than 150 GW of technical potential for hydropower generation, which is about double the capacity installed to date (WPC, 2019). Several states and cities (municipalities) in the US import Canadian hydropower to meet their climate change goals. The role of Canada's hydropower in contributing to reducing GHG emissions in North America increased when the US, Canada, and Mexico announced in 2016 a goal for North America to strive to achieve 50% clean power generation by 2025. Hydropower trade across the border also has a significant role to meet Europe's climate change and clean energy targets. The objective of this chapter is to explore and illustrate questions related to the power sector interconnection and trade in Europe and North America.

This chapter is organized as follows. Section 2 presents the historical and current status of power system integration in Europe and North America, followed by discussions of economics and investment models in Section 3. Section 4 presents a regulatory framework governing the power sector interconnection in Europe and North America. Issues and challenges faced by these integrated markets are highlighted in Section 5. Section 6 implies lessons that the South and Southeast Asia regions could learn from these markets, and Section 7 concludes.

2 Development of Cross-Border Electricity Interconnection and Trade

2.1 Power Market Interconnection Process

We start with a brief overview of the steps employed in Europe and North America to develop cross-border electricity interconnections and trade.

The purpose of presenting these steps is to indicate that the development of regional electricity trade takes time – many decades. The South and Southeast Asia regions could learn from the experiences of Europe and North America in developing an interconnected power system.

2.1.1 Europe

Cross-border electricity trade in Europe dates back to almost a century ago. It started in 1921 with the cross-border interconnection between France and Italy through Switzerland. The distance of the transmission line was 700 km (UCTE, 2003). Nothing changed over the following three decades. In 1951, eight countries in Western Europe established the Union for the Coordination of Production and Transmission of Electricity (UCPTE), which later became the largest synchronous interconnected system in the world covering connecting 16 European countries. The key objective of the UCPTE is to facilitate the optimal operation of member countries' power systems because the fuel economy was a big concern in the war devasted region.

By 1960, a uniform electricity grid with a 380 kV system extended across Western and Central Europe. The electricity grids of the eight founding members of UCPTE (Belgium, Germany, France, Italy, Luxembourg, Netherlands, Austria, and Switzerland) extended their grids eastwards, while Greece and Yugoslavia extended westwards, and Portugal and Spain joined them later. The regional electricity interconnection in Europe started with bilateral high-voltage transmission lines, which were then extended to multiple countries. This system allowed electricity trading between coal-based and hydro-based electricity generation regions. It also facilitated the seasonal trading of electricity and served as a backup for any major electricity system failure. By the early 1960s, Eastern European countries also established an interconnected electricity system, the Central Dispatch Organization of the Interconnected Power Systems (CDO/IPS). Electricity interconnection between the Western and Eastern blocs of European countries also started, but the interconnections were asynchronous and limited (Steinbacher et al., 2019). Under the UCPTE, countries established coordinated load dispatching by utilizing the telephone system. The coordinated dispatching allowed them to utilize excess capacity in a country in a given time to meet the load in other countries. It also facilitated the long-term and short-term power-purchasing contracts between multiple countries. By 1975, the European electricity transmission system has interconnected 32,200 MW of power generation capacities across the region.

Electricity sector liberalization and restructuring in Europe in the 1990s further facilitated the power system integration in the region (Jamasb & Pollitt, 2005). The EU introduced three consecutive legislative packages between 1996 and 2009 to integrate the electricity markets of its Member States. The 1996 Directive opened a window for an internal electricity market in the region by establishing common rules for the generation, transmission, and distribution of electricity for all its Member States and setting up a favorable ground for the expansion of cross-border electricity trade. In 2003, the European Commission (EC) established an advisory group, the European Regulators' Group for Electricity and Gas (ERGEG), to assist the Commission in consolidating a single EU market for electricity and gas. ERGEG launched the Electricity Regional Initiatives (RIs) in 2006 to accelerate the process of creating an integrated Energy Market (IEM). As an intermediate step, it created seven regional electricity markets before creating a single, competitive EU electricity market.

European national Transmission System Operators (TSOs) started to voluntarily cooperate in 2009 through Regional Security Coordination Initiatives (RSCIs). This is followed by the establishment of the European Network of Transmission System Operators for Electricity (ENTSO-E) in 2009. ENTSO's responsibilities include the development of policy positions, contributing to the development and implementation of common European network codes, facilitating technical cooperation between TSOs, developing long-term pan-European network plans, and coordinating R&D planning. It represents 43 TSOs from 36 European countries. Since 2018, the ENTSOs for electricity (ENTSO-E) and gas (ENTSO-G), have started to jointly develop these long-term scenarios (up to the year 2050) for a more comprehensive view on the requirements of the future European energy system. The TYNDPs are being prepared by the six ENTSO-E system development regional groups.

In 2011, the Agency for the Cooperation of Energy Regulators (ACER) was created that replaced the ERGEG which was established in 2003. ACER played a key role in regional cooperation for electricity market integration. It developed common network codes and set targets for switching from the regional markets, established earlier, to a single harmonization of the market on the EU level. In 2002, the European Council set a 10% electricity interconnection target (defined as import capacity over installed generation capacity in a Member State) by 2020. The target was revised in 2014 due to a lack of enough cross-border interconnection capacity to meet the target. The new target was set at 15% by 2030. In 2018, the Transmission System Operators and power exchangers started the Cross-Border Intraday initiative (XBID Project) to create a joint integrated intraday cross-border market, an important step toward integrating European electricity markets through market competition and pricing.

2.1.2 North America

The history of electricity cross-border trade between Canada and the US started even before the electricity system development in Canada. It dates back to 1901–1902 when American companies developed hydroelectricity projects on the Canadian side of the Niagara River through their Canadian subsidiaries. Most of the electricity generation from the projects was sold to the US (Martin-Nielsen, 2009). Since these early power plants were of greater benefit to the US than to Canada while exploiting Canadian resources, Canada regulated the electricity trade through the 1907 Act, which prohibited electricity sale to US consumers at a lower price than to Canadian consumers. This restriction was, however, never implemented.

Several long-term power trade agreements were signed between Canada and the US during the 1910–1915 period (Martin-Nielsen, 2009). Some of these contracts were the 85-year power trading agreements between Canada's Montreal Light, Heat and Power Company and New York in 1912; a long-term contract between Canadian Cedars Rapids Transmission Company and Aluminum Company of New York in 1912; and the contract between the Southern Canada Power Company of Québec and Vermont in 1914. Exports of electricity from Canada attracted regular debate on whether Canada should export its electricity to the US. Despite huge political pressure to ban electricity exports, the Canadian government did not ban them due to the threat of potential reprisal from the US. Instead, the Canadian government imposed a duty on Canadian electricity export in 1925.

Cross-border electricity interconnections between the Canadian province of Québec and the US (New York) started in 1912 when the Les Cedres (Cedars) generating station on St. Lawrence supplied electricity to the Alcoa smelter in Massena, NY. Hydro-Québec (Québec's electric utility) signed an 800 MW power export contract with New York's electricity authority in 1973. The construction led to 765 kV, 1,200 MW of cross-border transmission line in 1978. In 1985, Hydro-Québec operated 24-km long, 120 kV, 225 MW, transmission interconnection to supply electricity to Vermont. Hydro-Québec built a 172 km, 450 kV, 690 MW HVDC cross-border transmission link to connect to the New England Transmission system in 1986. The interconnection was further expanded later to transfer 2,000 MW of electric power. Hydro-Québec and the US company Central Maine Power are currently developing a 1,200 MW transmission interconnection project to supply clean hydropower from Québec to the New England Grid (to sell to Maine). In 1970, the Canadian province of Manitoba was interconnected to the US through a 230 kV transmission line with 375 MW power transmission capacity between Manitoba and North Dakota, followed by a second cross-border line with 230 kV, 250 MW in 1976. A third transmission line with 500 kV and 1,250 MW was commissioned connecting Manitoba to Minnesota in 1980 (Manitoba Hydro, 2013). In 2002, the fourth interconnection with 230 kV was constructed between Manitoba and North Dakota. The total power transfer capacity of these interconnections is 2,175 MW for export and 700 MW for import (Manitoba Hydro, 2013).

In Western Canada, all three provinces (Alberta, British Columbia, and Saskatchewan) are interconnected with the US grids. The Canadian province of British Columbia has become part of the interconnected system in Pacific Northwest since the signing of the Pacific Northwest Coordination Agreement in 1964 following the Columbia River Treaty. This agreement ensures close cooperation in the dispatching of major hydroelectric generating plants and electric systems that serve the Pacific Northwest. The latest interconnection was between the Canadian province of Alberta and the US state of Montana. The Montana–Alberta Tie Line (MATL), a 345 km, 230 kV, 300 MW AC transmission line connecting Lethbridge, Alberta, to Great Falls, Montana, was completed in 2013. It helps Alberta to reduce the carbon footprint of its fossil-fuel dominant electricity system by importing wind power from Montana.

Cross-border electricity interconnections and trade were facilitated with the establishment of the North American Reliability Council (NERC) in 1968 to ensure the reliability and adequacy of bulk power transmission across various electricity supply jurisdictions (or grids) in North America. The NERC's scope increased after the passage of the US National Energy Act of 1978. This Act allowed Canadian electricity grids to join the NERC. In the early 1980s, Canada introduced the National Energy Program, which was aimed at increasing Canadian hydrocarbon production, but later, it also facilitated electricity production for export to the US. After the 2003 electricity blackout in North America, the US and Canada harmonized the reliability standards and developed a single Electric Reliability Organization (ERO). More than 100 standards were developed to cover the reliability of resources and demand balancing, transmission operations, system planning, facility design and maintenance, cyber and physical security, communications and training, and preparedness for an emergency. In 2006, the NERC was certified as the ERO for the US; the NERC's regional entities served as the ERO for Canada. Since 2019, the National Energy Board has been designated for ERO with its new name, Canada Energy Regulator (CER). Hydro-Québec, the provincial electric utility of Québec, has been an active participant in the New York Power Pool since the 1980s; it has continued to actively participate in the New York Independent System Operator (NYISO), which is a successor of the New York Power Pool since 1999.

Although the electricity trade between the US and Mexico is small when compared to trade between the US and Canada, the history of the former is as old as that of the latter. In Mexico, the power sector is not fully privatized and electricity prices, mainly for residential customers, are heavily subsidized, which led to limited electricity trade with the US (CRS, 2017). The electricity system of Baja California, a Mexican province, has been interconnected with that of California for almost a century. The Mexican power system reform that was started in 1993 to introduce a competitive electricity market structure in the country increased the prospects for electricity trade with the US. The first asynchronous interconnection between ERCOT (Electricity Regulatory Council of Texas) and Mexican electric utility (Comisión Federal Electricidad (CFE)) was commissioned by American Electric Power Texas in Eagle Pass, Texas in 2001.

In 1989, Canada and the US signed the Canada–US Free Trade Agreements (CUFTA) to accelerate the US and Canada trade, including electricity. This agreement caused a significant increase in electricity trade between the two countries. It was replaced with the North American Free Trade Agreement (NAFTA) in 1994, a trade deal between Canada, Mexico, and the US. In 2018, these three countries signed a new Canada–US–Mexico Agreement (CUSMA) to replace NAFTA. In 2016, during the North American Leaders Summit, Canada, Mexico, and the US announced a goal of 50% clean power generation (e.g., hydro, nuclear, and other renewables) across the continent by 2025. This would certainly increase the cross-border flow of Canada's hydropower to help to meet the target.

There are several transmission interconnections projects in the pipeline across the borders of North America. These include (i) the 117 KM, 1,000 MW ITC Lake Erie underwater HVDC Connector transmission project across Lake Erie (between Nanticoke, Ontario, and Erie, Pennsylvania), to connect Ontario's power market with the 13 US states in the PJM power grid; (ii) the 240 km, 1,000 MW, underwater transmission interconnection through Lake Champlain to link between Québec and Vermont; (iii) the 385 km, 750 MW Great Northern Transmission Line between Manitoba and Minnesota; and (iv) the 257 km, 1,000 MW, Champlain Hudson Power Express connecting Québec to New York City.

2.2 Current Status of Regional Electricity Interconnection and Trade

2.2.1 Europe

The power system in Europe is extensively interconnected within the region and neighboring regions or countries. Europe has six sub-regions and each of the regions adopts a synchronous operating mode or operates as a single grid: Continental Europe, the Nordic, the Baltic, British, Ireland, and Northern Ireland areas. These areas are interconnected through AC or DC transmission links. The European interconnected markets are also linked with neighboring electricity markets, Ukraine and Turkey in the East, and North Africa in the South (Algeria, Morocco, and Tunisia).

The infrastructure of cross-border electricity interconnection in the EU has been steadily expanding over the past several decades. A regional institute ENTSO-E is established to govern the issues related to regional electricity trade in the EU. At present, electric power flows across the border between 36 European countries through 480,000 km of high voltage power network in 2018, more than the distance between the Earth and the Moon (ENTSO-E, 2019a). The total electricity exchange between 36 European countries is about 458.3 TWh of imports and 443.7 TWh of exports.

2.2.2 North America

2.2.2.1 US and Canada

Like the European electricity market, the North American electricity market, particularly that of Canada and the US, is highly interconnected. There are more than 30 major cross-border transmission interconnections between these two countries (Figure 7.2). The US and Canadian power systems are not only interconnected between the provinces of Canada and states in the US, but they are also the integrated part of the NERC, a not-for-profit North American organization responsible for ensuring reliability and adequacy of bulk power transmission in the electricity systems of North America (see Section 4 for more details).

It is interesting to note that the Canadian electricity system is more interconnected with the US electricity system than it is within Canada itself. There are four large-scale cross-border links between Canada and the US (Figure 7.2) and they are: (i) Canadian province of British Columbia to the Pacific Northwest electricity grid in the US; (ii) Canadian province of Manitoba to Midcontinent ISO of the US; (iii) Canadian provinces of Ontario and Québec to New York ISO of the US; and (iv) Canadian provinces of Québec to New England ISO of the US.

Several factors have played a role in the high-level integration between Canada and the US. The primary driver is the distance between the Canadian power plants and the US electricity markets. About 75% of the Canadian population lives nearby the US border. Economically, it is more attractive for Canadian electric utilities to serve their jurisdictions near the US border areas by extending their transmission networks to access the US markets, whereas connecting the eastern and western electricity grids within Canada would be a huge investment undertaking. Another important driver of Canada and the US cross-border electricity interconnection and trade is the existing regulatory framework in both countries. The electricity systems of the trading jurisdictions between Canada and the US are better harmonized technically than the eastern and western grids within Canada. Canadian provinces have joined the NERC, a single entity to govern the regulatory regime for its members in North America.

In Canada, provinces exercise full authority over their electricity systems. Their electric utilities, which still have a vertically integrated structure

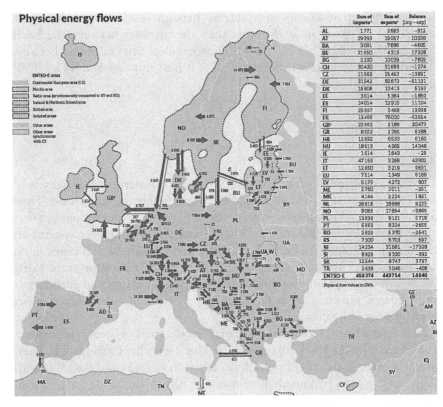

	Sum of imports[1]	Sum of exports[1]	Balance (imp−exp)
AL	1771	2683	−912
AT	29393	19057	10336
BA	3091	7696	−4605
BE	21650	4313	17338
BG	2220	10029	−7809
CH	30420	31693	−1274
CZ	11562	25453	−13891
DE	31542	82673	−51131
DK	15606	10413	5193
EE	3514	5364	−1850
ES	24014	12910	11104
FI	29397	9459	19938
FR	13466	76020	−62554
GB[2]	22652	2189	20473
GR	8552	2265	6288
HR	12692	6533	6160
HU	18613	4265	14348
IE	1614	1643	−29
IT	47169	3268	43902
LT	12850	3219	9631
LU	7514	1349	6166
LV	5179	4272	907
ME	2760	3011	−251
MK	4144	2224	1921
NL	26818	18596	8223
NO	8085	17954	−9869
PL	13839	8121	5718
PT	5669	8324	−2655
RO	2829	5370	−2541
RS	7300	6703	597
SE	14234	31561	−17328
SI	8928	9320	−392
SK	12544	8747	3797
TR	2638	3046	−408
ENTSO-E	458274	443734	14540

Physical flow values in GWh.

FIGURE 7.1 Electricity transmission networks in Europe. The arrows indicate the direction of electricity flow (GWh) between the borders.

Source: ENTSO-E (2019b)

Notes: AD (Andorra), AT (Austria), AL (Albania), BA (Bosnia and Herzegovina), BE (Belgium), BG (Bulgaria), CH (Switzerland), CY (Cyprus), CZ (Czech Republic), DE (Germany), DK (Denmark), EE (Estonia), ES (Spain), FI (Finland), FR (France), GB (United Kingdom), GR (Greece), HR (Croatia), HU (Hungary), IE (Ireland), IS (Iceland), IT (Italy), LT (Lithuania), LU (Luxembourg), LV (Latvia), ME (Montenegro), MK (Macedonia), NL (Netherlands), NO (Norway), PL (Poland), PT (Portugal), RO (Romania), RS (Serbia), SE (Sweden), SI (Slovenia), SK (Slovak), and TR (Turkey).

in hydropower-dominated provinces (British Columbia, Manitoba, and Québec), can make decisions regarding the markets for electricity production. On the US side too, states and ISOs are free to choose the sources of electricity to meet their demand within the existing rules and regulations. For example, consistent with climate change mitigation goals, California buys clean hydropower from the Canadian province of British Columbia instead of coal-based generation from nearby US states. One of the key drivers in the

FIGURE 7.2 North American interconnection map.

US for trading electricity with Canada is to get access to relatively cheaper and cleaner hydropower. Electricity imports from Canada have reduced the wholesale prices of supplying electricity by the US ISOs, for example, New England ISO (GTM, 2017).

2.2.2.2 US and Mexico

Compared to the cross-border transmission interconnection between the US and Canada, the cross-border interconnection between the US and Mexico are limited. The US states with transmission interconnections with Mexico are Arizona, California, New Mexico, and Texas. Unlike in Canada, these interconnections are asynchronous. Mexico shares the longest border with the ERCOT. The electricity system of Baja California in Mexico is connected to that of California in the US; however, the former has not connected to the rest of Mexico.

Compared to the US–Canada electricity trade, the US–Mexico electricity trade is limited. There are many reasons for this. The most important factor is the low load density between both sides of the borders. Due to the lack of adequate markets on both sides of the US–Mexico border, both electricity generation and transmission interconnection facilities are limited. The ongoing power sector reform in Mexico is expected to increase the electricity transmission interconnections capacity with the US and the cross-border electricity trade.

3 Economics and Investment Models

3.1 Economics of Regional Electricity Interconnection and Trade

There are multiple channels through which the benefits of cross-border electricity trade accrue. Some notable channels are (i) infra-marginal trade through the use of already built electricity generation capacity to generate export revenues especially when the capacities cannot follow the load quickly (e.g., steam turbine technologies); (ii) revenues through exports of electricity from the exploitation of yet untapped generation resources (e.g., hydropower, solar, wind, lignite, natural gas); (iii) reduction of supply cost of electricity to meet the demand through electricity imports; (iv) sharing of peak loads thereby saving the costs electricity system expansion to meet the peak loads; (v) sharing of reserve margins avoiding the cost of building electricity plants to meet the requirement of reserve margins; and (vi) helping to meet climate change mitigation targets through the trade of non-CO_2 electricity.

Some analysis presents the impacts of cross-border electricity trade in general where aggregate benefits are assessed without identifying the specific channel identified above. Some other studies focus on a specific window through which the benefits could be realized. We first discuss some studies that estimate the overall benefits of cross-border electricity trade in the EU and North America. We then discuss studies that estimate the benefits of regional electricity trade from the perspective of inframarginal trade and climate change mitigation.

3.1.1 Economic Impacts in General

Using economic modeling, some studies have estimated the economic impacts of various levels of integration of power systems in Europe. For example, a study by Booz and Company et al. (2013) estimates that the economic benefits of implementing the Target Electricity Model (TEM) that the EC announced in 2004 are in the range of 12.5 billion Euro to 40 billion Euro per year by 2030. Using the European Climate Foundation modeling tool, Zachmann (2013) estimates that integration of electricity markets in Europe could save up to 426 billion Euro by 2030.

In North America, the total value of electricity trade between Canada and the US varies from US$1.9 billion to about US$3.2 billion during 2010–2019. The value of Canadian electricity exports to the US is much higher than US electricity exports to Canada. In fact, the US has provided market opportunities for Canada's hydropower. Also, Canadian hydropower exports to the US contributed to reducing CO_2 emissions in the US. Both countries have gained from the cross-border electricity trade.

3.1.2 Economic Impacts through a Specific Channel

3.1.2.1 Benefits through Inframarginal Trade

The infra-marginal trading opportunities occur because of typical characteristics of electricity to flow either side of the border in the presence of cross-border transmission lines. Two electricity grids might have different load profiles and at the margin, they might be running power plants with substantially different costs (or they have different prices). This difference in costs or prices between grids provides an opportunity to trade and benefits both parties. The exporting party gets revenue from electricity that would not have been generated otherwise. The importing party benefits because it gets the supply cheaper than it would have been otherwise. If there were no cross-border interconnection between Canada and the US, the Canadian hydropower plants would lose their revenues from electricity sales to the US.

The cross-border electricity transmission interconnections between the US and Mexico provide opportunities for the US generators to export electricity to Mexico at higher prices, whereas at the same time Mexican industrial consumers benefit as they get electricity from the US suppliers at lower prices compared to what they could get otherwise. This is because there is a substantial difference between the marginal costs of electricity between the two borders. In the case of Canada and the US, the infra-marginal electricity trade is governed by interruptible electricity trade agreements where market participants exchange electricity between the borders as dictated by the market situations. In 2019, Canada exported 60,376 GWh of electricity, thereby generating about US$2 billion in revenues. On the other hand, power producers/traders in the US exported 13,369 GWh and earned US$446 million in revenue (NEB, 2020).

Analyzing the benefits of further integrating EU electricity markets through the TEM, Newbery et al. (2016) find that following the TEM objectives would deliver aggregated benefits for the region that is much higher than the costs needed to realize the integration under TEM. The potential benefits amount to 3.9 billion Euro annually. They also estimated the shares of various elements in the total benefits, e.g., the share of benefits from day-ahead markets is 26%, balancing markets is 35%, unscheduled flow is 35%, curtailment is 3%, and intraday coupling is 1%.

Cross-border interconnection has provided a unique opportunity for some provinces as it facilitates seasonal trading of renewable electricity which varies significantly across the seasons. For example, the Canadian Province of British Columbia often gains from such an opportunity.

3.1.2.2 Climate Change Mitigation Benefits

Several studies have estimated the economic benefits of regional electricity trade through non-CO_2 electricity trade to help meet climate change commitments in the EU and North America. For example, Abrell and Rausch (2016) investigate a policy of expanding cross-border transmission capacities for electricity in Europe under the TYNDP, which is the main vehicle to expand cross-border electricity transmission capacity in Europe. Using a hybrid modeling system that linked a multi-sector, multi-country computable general equilibrium model with an energy system model to estimate the benefits, they find that the enhanced electricity trade through execution of the TYNDP project could benefit the EU from US$1.6 billion to US$2.6 billion (2011 price) annually. If the EU countries go beyond the TYNDP and exercise unrestricted electricity trade across the borders, the aggregated gain (EU-wide) would be US$5.8 billion to US$8.7 billion (2011 price) annually. There are other economic benefits too. For example, Antweiler (2016) developed an innovative method to estimate the benefits of cross-border electricity trade between Canada and the US through "reciprocal load smoothing". It refers to smoothening electricity load curves of trading countries and is equivalent to subsidizing the rent-seeking entry of oligopolistic exporters into each other's markets. It is a new type of gain from electricity trade considering the electricity market's characteristic of stochastic demand variation and strongly convex (upward-sloping) marginal costs.

The importance of cross-border transmission interconnections has also increased due to their potential roles of providing markets to variable renewable energy (VRE) sources that help meet EU countries' climate change commitments. Northwestern Europe has abundant flexible hydropower; for instance, countries like Denmark possess a good potential for wind power, both onshore and offshore type. Using a standard electricity system optimization model, Chen et al. (2020) quantify the economic benefits of cross-border transmission to help decarbonize the electricity supply systems in Northwestern Europe by 2050. They also estimated the benefits of cross-border interconnection investment and trade, and show that 76 GW of cross-border transmission capacity, in addition to that in the reference scenario, would be added during the 2030–2050 period. This addition is four times as high as the planned expansion by 2030. The total cost of electricity system expansion would be 5% smaller than that in the reference scenario. The trade would, however, drop electricity prices in western Europe by 6%

at the cost of northern European consumers where electricity price increases by 21%. This type of asymmetry in benefits sharing always puts hurdles on the expansion of cross-border transmission interconnection and trade. If the external (or social) benefits are also accounted for, the trade would reduce 40% of emissions from the reference scenario in 2050 in the trading jurisdiction.

3.2 Investment Models for Cross-Border Interconnections

Different types of models have been exercised to fund cross-border transmission interconnections. Having a Parliament at the EU level that is interpreted as the government for the EU, transmission interconnections are funded mostly by the EU central funds. In North America, the funding mechanisms are different ranging from transmission interconnections funded and owned by state-owned electric utilities to privately owned cross-border transmission interconnections. Below, we discuss the investment models exercised in the EU, followed by those in North America.

3.2.1 Investment Models in the EU

The most common funding or investment model for cross-border transmission interconnections in the EU is the EU central funding. Where there could be several potential cross-border projects, they are prioritized based on their needs and benefits. Some interconnections are urgently needed to avoid transmission congestion, whereas other interconnections are developed based on long-term benefits from economic as well as environmental perspectives. The Trans European Network for Energy (TEN-E), a framework to facilitate the linking of the energy infrastructure of the EU Member States, provides financing to identify strategic European transmission interconnection projects based on the overall costs and benefits of the projects.

The EU also provides further financial supports for the identified projects as 'projects with common interest' through multiple instruments, such as the Connecting Europe Facility (CEF). This is a key EU funding instrument to promote growth, jobs, and competitiveness through targeted infrastructure, including cross-border electricity transmission lines. The Innovation and Networks Executive Agency (INEA) is responsible for implementing the CEF programs. In the 2014–2020 budget, 5.12 billion Euro of CEF funding was allocated for trans-European energy infrastructure projects, it has been increased to 8.7 billion Euro in the 2021–2027 budget program (EC, 2019).

In terms of ownership of cross-border transmission lines in the EU, two models can be observed: TSO-owned and merchant-owned. The former is the model where the transmission interconnections are built and owned by the TSOs, the latter is built and owned by the private sector. In the TSO-ownership model, TSOs are responsible for the development and operation of cross-border transmission interconnections. The costs of building and operation are reimbursed through the regulated network charges. In the merchant ownership model, independent private entities develop and operate cross-border transmission interconnection assets. The revenues to recover the investments are regulated through a cap and floor mechanism, where the revenues are capped at the level to ensure the minimum ROI based on 25 years of the financial life of the project (Steinbacher et al., 2019).

3.2.2 Investment Models in North America

The investment model for cross-border electricity connection in North America is different from that of the EU as there is no single entity to centrally fund an interconnection project, as the EC in the case of the EU. Most of the financing and owning of cross-border transmission lines in North America are by private companies like in the case of generation assets (merchant model). In some cases, especially in those jurisdictions where state-owned utilities exercise the vertical monopoly type of utility structure (e.g., Hydro-Québec, Manitoba Hydro), they build and own, jointly with their US partners, the cross-border transmission lines. Intuitively, the Canadian segment of the transmission line is built and owned by Canadian utilities and the US segment of the transmission line is built and owned by US companies. Governments, federal and provincial/state, also contribute to the financing of cross-border transmission lines.

Merchant investment model is also used to finance cross-border transmission facilities exercised in North America. The MATL transmission interconnection project is an example. The Canadian segment of the 345 km long, 230 kV, 300 MW MALT project was financed by a subsidiary of a Canadian energy company, Enbridge. It invested US$300 million out of 400 million total costs to complete the Canadian segment (123 km) of the transmission interconnection. This project could also serve as an example of US federal funding for cross-border transmission interconnection projects. The Western Area Power Administration (WAPA) of the US received US$3.25 billion for various projects under the federal stimulus package that the US federal government provided in response to the 2008 financial crisis. MATL transmission interconnection project also received funding from this stimulus package.

4 Regulatory Frameworks for Cross-Border Electricity Interconnections and Trade

Cross-border interconnections and trade are governed by the existing rules and regulations of the trading partners or the trading region where regional institutions are created for this purpose. In the EU, it has set up several rules, regulations, guidelines, and frameworks for cross-border electricity inter-connection and trade. In North America, cross-border trade is governed on a bilateral basis, between the trading governments.

4.1 Electricity Regulatory Legislative and Entities in the EU

In Europe, particularly, in the EU region, cross-border electricity intercon-nections and trade are largely governed by rules and regulations set by the European Parliament. Over the past three decades, the EU has adopted a series of directives, known as Energy Packages, for the liberalization and the harmonization of energy (electricity and gas) markets. The most recent directive, the Fourth Energy Package, also known as the *Clean Energy for All Europeans* (CEP) package, consists of Electricity Market Directive (2019/944/EU) and three regulations. The three regulations are the Electricity Regulation (2019/943/EU), the Risk-Preparedness Regulation (2019/941/EU), and the Agency for the Cooperation of Energy Regulators (ACER) Regulation (2019/942/EU) (EU, 2020). The Fourth Energy Package sets a legal framework to facilitate the achievement of 2030 energy and cli-mate targets in the EU.

At the EU level, cooperation for common electricity markets facilitates the coordination between different entities, including the Member States, their NRAs, TSOs, and prospectively Distribution System Operators (DSOs). In the EU, electricity generation is separated from the transmission system operation. The transmission grids are operated by national TSOs, and the distribution networks are managed by DSOs. The TSOs are entities operating independently from electricity generation and distribution com-panies and they are responsible for ensuring a supply of electricity meets the demand at each instant of time. TSOs are either wholly or partly owned by state or national governments. The legislative processes (e.g., Electricity Regulation or Electricity Directive), the subordinated technical regulations (EU Network Codes), and the common decisions on terms, conditions, and methodologies by the NRAs are key to integrated EU electricity market. To enhance the cooperation between NRAs, the EU mandated its Member States to establish NRAs and created the ACER. The ACER is established in March 2011 by the Third Energy Package legislation as an independent body to foster the integration and completion of the European IEM both for electricity and natural gas (ACER, 2019). ACER fosters a fully integrated and well-functioning IEM, where electricity and gas are traded and supplied according to the highest integrity and transparency standards, and the EU consumers benefit from a wider choice, fair prices, and greater protection. The four main objectives of the agency are (i) to contribute to the completion of the IEM and monitor its functioning, (ii) to contribute to the infrastruc-ture challenges, (iii) to increase the integrity and transparency of wholesale energy markets, and (iv) to contribute to addressing longer-term regulatory challenges.

ACER also monitors the work of ENTSO-E and its EU-wide network development plans as well as the functioning of the common energy markets. The agency plays a role as an independent facilitator for regional cooperation in the field of electricity market integration and oversees the supervision of regional cooperation of TSOs. To fulfill its mission, ACER can issue non-binding opinions and recommendations to national energy regulators, TSOs, and EU institutions. In areas defined within European legislation, ACER can take binding individual decisions in specific cases and under certain conditions on cross-border infrastructure issues. ACER, together with ENTSO-E, also develops draft framework guidelines that serve as a basis for the drafting of Network Codes upon the request of the EC (Steinbacher et al., 2019).

Likewise, ENTSO-E plays an important role in the cross-border electricity markets and contributes to the design and implementation of market rules. Mandated by the EU's Third Legislative Package for the IEM, ENTSO-E is established in 2009 to promote cooperation across Europe's TSOs and to support the implementation of EU energy policy. The ENTSO-E is the successor of six regional associations of electricity system operators. The main objective of ENTSO-E is to set up the IEM and ensure its optimal functioning, and of supporting the ambitious European energy and climate agenda. One of the important issues in the EU agenda is the integration of a high degree of Renewables in Europe's energy system, the development of consecutive flexibility, and a much more customer-centric approach than in the past. The ENTSO-E is also committed to developing the most suitable responses to the challenge of a changing power system while maintaining the security of supply. Innovation, a market-based approach, customer focus, stakeholder focus, security of supply, flexibility, and regional cooperation are key to ENTSO-E's agenda. The ENTSO-E contributes to the achievement of these objectives mainly through (i) policy positions, (ii) drafting of network codes and contributing to their implementation, (iii) regional cooperation through the RSCIs, (iv) technical cooperation between TSOs, (v) publication of Summer and Winter Outlook reports for electricity generation for the short-term system adequacy overview, (vi) development of long-term pan-European network plans, and (vii) coordination of R&D plans, innovation activities and the participation in Research programs like Horizon 2020, formerly known as 7th Framework Programme (FP 7). The ENTSO-E operates as a non-profit organization and is financed by its members.

The Council of European Energy Regulators (CEER) is another council that was established in 2020 for the cooperation of the independent energy regulators of Europe. The council is the voice of Europe's NRAs at the European and global levels. The CEER's 39 members and observers are the independent statutory national energy regulatory authorities from across Europe. CEER is legally established as a not-for-profit association under Belgian law. Through CEER, national regulators cooperate at the EU level and speak with one voice globally. The CEER supports its NRA members and observers in their daily responsibilities, sharing experience, and developing regulatory capacity and best practices. It does so by facilitating expert working group meetings, hosting workshops and events, supporting the development of regulatory papers, and through our in-house Training Academy. In terms of the policy, CEER actively promotes an investment-friendly, harmonized regulatory environment, and the consistent application of existing EU legislation. A key objective of CEER is to facilitate the creation of a single, competitive, efficient, and sustainable IEM that works in the public interest (CEER, 2020). The CCER's work complements the work of the ACER and they both share similar objectives.

European DSOs (E.DSO) are the key interface between Europe's DSOs and the European institutions and stakeholders. The main purpose of E.DSO is to promote the development and large-scale testing of smart grid technologies in real-life situations, new market designs, and regulation. E.DSO gathers 41 leading electricity DSOs in 24 countries, including two national associations, cooperating to ensure the reliability of Europe's electricity supply for consumers and enabling their active participation in the energy system by shaping smarter grids for your future (E.DSO, 2020). E.DSO focuses on guiding EU research, demonstration and innovation, policy, and Member State regulation to support smart grid development for a sustainable energy system.

The network codes drafted by the ENTSO-E are enacted by the EC considering the view of the Member States and the European Parliament. Following their adoption, these network codes are European regulations that are legally applicable in each EU Member States.

4.2 Electricity Regulatory Regime in North America

The cross-border electricity system across North America is quite different from the EU. In North America, the cross-border electricity system is well integrated between the US and Canada, while it is less integrated between the US and Mexico. Realizing the mutual benefits of cross-border electricity trade, such as strengthening the security and resilience of an integrated cross-border electricity grid, providing increasing amounts of clean energy, and improving economic competitiveness in the region, leaders of these trinational countries have affirmed their support for the concept of increasing integration of the electricity system in the region.

The aftermath of the two large-scale cross-border electricity blackouts, the Great Northeast Blackout of 1965 and the Northeast Blackout of 2003, led to the creation of the NERC. The NERC is a non-profit ERO for North America established in 2006. The ERO is comprised of the NERC and the six regional entities. The NERC's mission is to assure the effective and efficient reduction of risks to the reliability and security of the grid (NERC, 2020). The NERC is responsible for developing and enforcing the reliability standards, annually assessing the seasonal and long-term reliability, monitoring the bulk power system through system awareness, and educating and training industry personnel. The NERC's area of responsibility spans the continental US, Canada, and a part of Mexico (the northern portion of Baja California). The NERC is subject to oversight by the Federal Energy Regulatory Commission (FERC) of the US and the governmental authorities in Canada. The NERC's jurisdiction includes users, owners, and operators of the bulk electricity system, which serves more than 400 million people. The NERC's role in Canada similar to the US. While the process for approving the NERC Reliability Standards varies in the different Canadian jurisdictions, standards are mandatory and enforceable in six of the ten provinces (i.e., British Columbia, Alberta, Saskatchewan, Manitoba, Ontario, Québec, New Brunswick, and Nova Scotia). Enforcement programs vary among the provinces, with provincial regulators having ultimate authority for monitoring and enforcing compliance in most provinces. However, authority over electricity generation and transmission in Canada rests primarily with provincial governments. However, the electricity regulatory entities in Mexico are quite different. Mexico enacted significant energy reforms in 2013 and 2014 that include restructuring the monopolistic Mexican electricity industry and increasing the opportunity for private investment and a competitive electricity market. The main three electricity authorities in Mexico are the federal energy regulator (Comisión Reguladora de Energia (CRE)), the independent system and market operator for all parts of the Mexican electric system (Centro Nacional de Control de Energía (CENACE)), and the government-owned utility (CFE). Because of the growing cross-border electricity-related operations between the US and Mexico, both governments are actively promoting the reliability and security of the interconnected electricity system through the NERC.

In many cases, cross-border integration of the electricity sector is best supported by the development of regional regulatory institutions, which work with local regulatory institutions. In general, the North American electricity system is heterogeneous. The operations and planning primarily take place through regional entities, and every part of the system has evolved with different characteristics and structures.

The cooperation for electricity integration between these three countries is strategically at the presidential and ministry levels, and technically at the agency level, although progress on some strategic efforts has been limited (GAO, 2018). At the presidential level, trilateral cooperation has occurred

TABLE 7.1 A summary overview of Canadian, US, and Mexican power sector structures and institutions

National/ subnational government (s)	NERC Balancing region	Market Operator/RTO /ISO	Ministry/ Agencies involved	Regulator	Market Design	Capacity in MW
Federal	--	--	Natural Resources Canada (NRCan)	National Energy Board (NEB)	--	135,000
BC	WECC	BC Hydro, participates in NWPP	Energy and Mines	BC Utilities Commission	Centrally managed model with bilateral contracts	15,220
AL	WECC	Alberta Electric System Operator, participates in NWPP	Department of Energy	Alberta Utilities Commission	Mandatory power pool	12,298
SK	MRO	SaskPower	Provincial Cabinet and Crown Investments Corporation of Saskatchewan	Saskatchewan Rate Review Panel	Centrally managed model with bilateral contracts	4,042
MB	MRO	Manitoba Hydro	Innovation, Energy and Mines	Public Utilities Board	Centrally managed model with bilateral contracts	5,640
ON	NPCC	Independent Electricity System Operator	Energy	Ontario Utilities Board	Power pool for real-time energy market with bilateral contracts, PPAs, and regulated tariffs	34,276
QC	NPCC	HQ	Natural Resources and Wildlife	Régie de l'énergie	Centrally managed model with bilateral contracts	42,485
NB	NPCC	New Brunswick System Operator	Department of Energy	Energy and Utilities Board	Physical bilateral market with a redispatch market	4,625

CANADA (border provinces)

(Continued)

TABLE 7.1 (Continued)

National/ subnational government (s)	NERC Balancing region	Market Operator/RTO /ISO	Ministry/ Agencies involved	Regulator	Market Design	Capacity in MW
Federal	--	--	DOE	FERC, EPA, BLM, NRC	--	1,063,000
WA, MT, ID, OR, WY,* CA,* NV, UT	Western Electricity Coordinating Council (WECC)	Control area operators: "Northwest Electric Markets"	Bonneville Power Administration (BPA); Western Area Power Administration (WAPA)	State regulatory commissions;	Traditional wholesale electricity markets	78,964
MN, MI, MT, IA, IL, ND, SD, WI, MI* AK, KY,* IN,* LA, MS, TX	Midwest Reliability Organization (MRO)	Midcontinent ISO (MISO)	WAPA	State regulatory commissions;	RTO/ISO competitive markets	180,711
NY	Northeast Power Coordinating Council (NPCC)	New York ISO (NYISO)		New York State Public Service Commission (PSC)	RTO/ISO competitive markets	30,039
VT, NH, ME, MA, CT, RI	NPCC	New England ISO (ISO-NE)		State regulatory commissions;	RTO/ISO competitive markets	31,000
OH,* IL,* KY,* WVA, MD, PA, DE, IN, NJ, NC, TN,* VA, DC	Reliability First (RF)	PJM	Southeastern Power Administration (SEPA)	State regulatory commissions;	RTO/ISO competitive markets	183,604
FL, GA, AL, MS, NC, SC, MO, TN	FRCC, SERC	Control area operators: "Southeast Electric Power Markets"	Tennessee Valley Authority; SEPA	State regulatory commissions;	Traditional wholesale electricity markets	238,000
AZ, NM, NV, WY, SD, CO, TX	WECC	Control area operators: "Southwest Electric Markets";	WAPA	State regulatory commissions;	Wholesale electricity markets	50,000

UNITED STATES

	Southwest Power Pool, RE (SPP)	Southwest Power Pool (SPP)	*Southwest Power Administration (SWPA)*	State regulatory commissions;	RTO/ISO competitive markets	
MT,* ND,* SD,* WY,* NB, IA,* KA, OK, TX,* NM,* AR, MO,* LA*	Southwest Power Pool, RE (SPP)	Southwest Power Pool (SPP)	*SWPA*	State regulatory commissions;	RTO/ISO competitive markets	78,953
TX	Texas Reliability Entity	Electricity Reliability Council of Texas (ERCOT)	*SWPA; WAPA*	Public Utility Commission of Texas	RTO/ISO competitive markets	75,964
CA, NV*	WECC	California ISO (CAISO)	*WAPA*	California Public Utilities Commission (PUC)	RTO/ISO competitive markets	60,000
Federal	*(none)*	CENACE (wholesale); CFE (residential users)	SENER, SEMARNAT	CRE	Wholesale market for industrial users; regulated market for residential	62,000
MEXICO — Baja California	WECC					2,341

Source: DOE (2016). Available online at www.energy.gov/sites/prod/files/2017/01/f34/Electricity%20in%20North%20America%20Baseline%20and%20Literature%20Review.pdf

Note: The US states listed with an asterisk (*) have only minor geographic areas included in the listed jurisdiction. Power marketing administrations (PMAs) are indicated in the Ministry/Agencies involved section in italics. Acronyms for Canadian provinces used are British Columbia (BC), Alberta (AL), Saskatchewan (SK), Manitoba (MB), Ontario (ON), Québec (QC), New Brunswick (NB), and Newfoundland and Labrador (NL). FRCC stands for Florida Reliability Coordinating Council; SERC stands for Southeastern Electric Reliability Council.

mainly through the North American Leaders' Summit, where the leaders of the three countries discuss economic issues, including electricity. At the ministerial level, the US secretaries of Energy and State hold trilateral and bilateral meetings with the Canadian and Mexican counterparts. At the agency level, agencies from these three countries collaborate on technical activities such as trade and regulatory issues. The basic entities of the North American electricity system include interconnections, reliability regions, balancing authority areas, utilities, RTOs/ISOs, and Power Marketing Administrations (PMAs). A summary overview of Canadian, US, and Mexican power sector structures and institutions is provided in Table 7.1.

5 Challenges and Concerns of Cross-Border Electricity Integration

Trading electricity across international borders can reduce costs, improve system reliability, and reduce emissions for all participating nations or regions. Despite these benefits, the value and quantity of worldwide cross-border electricity trade are negligible. For example, the worldwide total merchandise trade is about 38.7 trillion US$ in 2018, of which only 68.7 billion US$ accounts for cross-border electricity trade, an equivalent of 0.18% of total trade (UN, 2020). Likewise, only about 5.7% of total worldwide electricity production of 25,721 TWh was traded across international borders in 2017 (IEA, 2020). Region-wise, the bulk of this cross-border electricity trade occurs in Europe and North America.

The cross-border electricity system integration involves unique challenges. Broadly, these challenges can be grouped into economic, security, societal, and legislative/regulatory challenges. These challenges are not unique, but they are interrelated to each other. For example, economic, security, and societal challenges are linked to environmental challenges, while legislative or regulatory challenges encompass all challenges.

5.1 Economic Challenges

The economic benefits for all participating countries in the cross-border electricity trade and interconnection infrastructure come from lower investment costs, e.g., building cross-border or local interconnection infrastructure, due to increased economies of scale, and lower overall operating costs, due to increased system efficiency. However, the harmonization of decisions of participating countries on cross-border electricity trade issues, such as cost recovery, investment risk, and the distribution of economic benefits, is quite complex and challenging.

In Europe, one key economic challenge is the current low level of price convergence in most of the regions due to the limited amount of offered cross-border capacity and/or lack of market coupling (ACER & CEER,

2019). Only the Baltic region has seen a high level of full-price convergence, while other regions have seen a low level of price convergence. The differences in national electricity systems and the growing influence of renewable generation are mainly responsible for persistent price differences. Despite the extension of market coupling between borders, the other economic challenge is the level of efficiency in the use of interconnectors has remained unchanged in the past few years. For example, the level of economic efficiency in the use of interconnectors in the DA market timeframe increased from about 60% in 2010 to 86% in 2016, while it increased by only 1% to 87% in 2018 (ACCER & CEER, 2019).

Significant investment and costs are other important challenges for developing cross-border electricity exchange capacities. These investments and costs include deploying new transmission assets, reinforcing the surrounding grids, and maintaining the infrastructure. For example, in Europe, it is estimated that the investment needed for electricity transmission infrastructure ranges from 125 to 148 billion Euro by 2030 and from 300 to 420 billion Euro by 2050, depending on different scenarios (EC, 2017). Apart from market coupling and investment costs challenges, the information on the network tariff data in the EU is currently very heterogeneous at the Member State level and limited to a few types of generators and loads at the European level (EC, 2017). Transmission network tariffs are typically used to recover the fixed capital and operating (infrastructure) costs of providing the transmission network. It is also used to recover the costs of connecting new users (generation and load) to the network. Network tariffs are levied by the TSOs and DSOs and tariffs are regulated by the EU.

In North America, mainly between the US and Canada, there are concerns from generators in the US that increasing cross-border electricity trade would hurt domestic markets and give Canadian suppliers market power (DOE, 2017). There is also a large difference in average wholesale prices, mainly between the US and Mexico hindering the market coupling between them. For example, in 2018, average wholesale prices in the Baja California Interconnected System (Mexico) are US$49.8 per MWh (NERA, 2020), while in Texas (US), the ERCOT North 345 KV Peak wholesale price is much lower at US$30.2 per MWh (EIA, 2020).

5.2 Security of Supply Challenges

Unscheduled electricity flows, such as loop and transit flows,[3] often lead to insecure cross-border electricity trade. These effects cause external costs, e.g., costs related to the security of supply and reduced capacity to participating countries. In recent years, the unscheduled electricity flow is a growing concern for cross-border electricity trade in the European regions with higher shares of renewables that tend to vary with changing weather patterns.

The other security challenge is the insufficient availability of cross-border interconnections capacity. In the EU, there is currently not enough transmission capacity for cross-border electricity interconnections. For example, between 2016 and 2018, the average capacity made available for cross-zonal trade across the countries in Europe is much lower than the 70%, target set by the Electricity Regulation under CEP legislation (ACER & CEER, 2019). The relatively low values of the available cross-zonal capacities lead to network congestion, which is not addressed well by the existing bidding-zone configuration. Rumpf (2020) reports that restriction of cross-zonal capacity to relieve network congestion inside the domestic grids (internal constraints) by TSOs is pushing congestion to the border. For example, 72% of the time, the congestion at the Central West Europe flow-based market coupling is attributed to internal constraints.

In North America, the mode of cross-border integration is bilateral. The cross-border electricity trade between the US and Canada is much higher compared to the similar trade between the US and Mexico. For example, in 2018, the total quantity of electricity traded between the US and Canada is 149.2 TWh, equivalent to 5.3 billion US$, while it is only 15.4 TWh, equivalent to about 440.3 million US$, between the US and Mexico (UN, 2020). The low level of integration between the US and Mexico is attributed to the differing legal instruments for open-access transmission agreements and reliability coordination between these two countries (Krupnick et al., 2016). The complexity of several transmission projects also raises a variety of stakeholder concerns that lead to long development times and unexpected delays, such as environmental impacts of transmission infrastructure and potential implications of greater import of electricity from Canada to local and regional economic development in the US (DOE, 2017).

5.3 Legislative and Regulatory Challenges

Effectively managing the complex cross-border electricity system is a major challenge for regulators. The regulators' job is to ensure that the electricity market operates as efficiently and sustainably as possible and in the best interests of their consumers. In the EU, Energy Packages, and recently introduced CEP Package, and the European Network Codes, are some examples of the EU legal frameworks (regulations and directives) used for facilitating the cross-border electricity market. One key challenge in Europe is the asymmetric information between regulators and TSOs. TSOs may be given stronger financial incentives to maximize the capacity which is made available for commercial purposes. Another challenge is dealing with existing long-term contracts that limit TSOs' ability to utilize the interconnector capacity for market coupling. Also, the bureaucratic parts of REMIT,[4] such as registration, transactions, and orders reporting, and obligations to publish inside information, are quite complicated and challenging.

In North America, the US, Canada, and Mexico set a goal to achieve 50% clean power generation by 2025 through greater cross-border integration. However, a study by Krupnick et al. (2016) suggests that the complexity and current asymmetry of national and sub-national policy frameworks in these countries may impede achieving the goal. The siting and permitting decisions are made at the state and local level in the US, including for cross-border transmission networks, which is another challenging issue for the development of new cross-border transmission infrastructure (DOE, 2017). Also, the regulation of the transmission system and appropriate cost allocation is quite different in the US and Canada. In Canada, regulations of the transmission are made entirely at the provincial level, and the provinces negotiate the appropriate cost allocation of the construction of transmission, while in the US, FERC and the local authorities are involved (IEA, 2019).

5.4 Societal Challenges

Many cross-border electricity transmission infrastructure projects also have important public acceptance problems, such as perceived risks to health or intrusiveness of infrastructure in the landscape and impact on local climate and nature (EC, 2017). There are also considerable differences in terms of the energy mix, size of the electricity market, geographical location, and governance across participating countries that influences the cross-border electricity interconnections. If these societal and diverse participating countries' challenges are not addressed on time, the implementation of cross-border electricity interconnection projects often gets delayed causing additional investment costs.

5.5 Project Implementation Challenges

Cross-border electricity went through many challenges in the past and possesses several at present. It faces concerns from both sides. Many Canadians consider that their clean hydropower is being exploited to meet the US demand at very low prices, especially through earlier long-term contracts. On the US side, there are concerns from generators that increasing cross-border trade would give Canadian suppliers market power as the cheaper Canadian electricity could lower prices for US customers, thereby making US generators using fossil fuels less cost-competitive (i.e., losing market share).

One big challenge to enhancing the regional electricity trade is project cancellation and delays due to opposition from the public, political parties, and judiciary interventions. Investment in transmission lines connecting two borders is further risky because it must go through a series of regulatory hurdles – international, national, and local social (Huang & Van Hertem, 2018). Several proposed cross-border transmission projects in the Midwest and Northeast did not go through due to concerns raised by a variety of stakeholders, and long delays caused by the regulatory hurdles. One example is the Northern Pass cross-border transmission project to connect Québec to the New England grid. The project was jointly proposed by Québec's state utility Hydro-Québec, and American utilities Northeast Utilities and NSTAR in 2008. Hydro-Québec was planning to build and own the Canadian segment of the transmission line, whereas the US segment was to be built and owned by the Northern Pass Transmission LLC jointly owned by the two US companies, Northeast Utilities and NSTAR. The 300 km, 1,200 MW HVDC transmission line was estimated to cost around US$1 billion. The project was initially scheduled to be completed by 2015, later the completion schedule was delayed by five years. The purpose of the transmission line was to deliver clean hydropower from Québec to Massachusetts, passing through New Hampshire. It received stiff public protest in New Hampshire as they viewed it as a threat to the rural economy based on tourism. The project state Site Evaluation Committee of New Hampshire disapproved of the project in 2018. The developer, Eversource Energy, went to the New Hampshire Supreme Court, where it lost the case in 2019. Thus, after struggling for survival for 10 years with an expenditure of more than US$300 million, the project got canceled.

5.6 Electricity Market Challenges

The market is the main driver of the business. This is true for cross-border electricity trade. Since there is a high electrical load density between both sides of the US and Canada border, the electricity trade is prosperous between them. However, electricity markets are thin on either side of the US and Mexico border because of low population density. The states in Mexico along the US border have some of the lowest population densities in the country. Low population density means low electricity load density or low electricity demand. Therefore, there exists only a limited cross-border electricity trade between the US and Mexico. In the absence of market or demand, investing in cross-border transmission interconnection facilities is a waste of resources. Therefore, the presence of a strong market is the precondition for cross-border electricity trade.

6 Lessons Learn from the Experience of the EU and North America

Both Europe and North America have gone through more than 100 years of experience in cross-border electricity interconnection and trade. Their rich experience can provide very useful lessons for many other regions in the world that have already initiated regional electricity interconnection and trade or planning to do so. Here, we highlight some of the key lessons that India's North Eastern Region (NER), South and Southeast Asia could learn from the European and North American experiences.

6.1 Long Journey Requiring Persistent Efforts

Since cross-border electricity cooperation and trade involve multiple governments, the path is not straightforward. The process of cooperation could be lengthy and could face resistance in some form from the participating countries. Between Canada and the US, the electricity trade started even before the development of power systems in the respective countries. When American companies, through their Canadian Subsidiaries, developed power projects on the Canadian side of Niagara Falls and brought all the generated power to the US in the very beginning (before 1907), there was a natural disappointment on the Canadian side. Canadians did not have an electricity supply, whereas Americans were using their natural resources. Therefore, the Canadian government introduced the 1907 Act to regulate the power trade and to force power producers to increase their supply of electricity in Canada. There was a strong view of the people and even the political section that Canada should stop exporting electricity to the US. Some Canadian governments, such as the governments of Prime Minister Mackenzie King, were under public pressure to ban Canadian electricity export to the US (Martin-Nielsen, 2009). Despite the upheaval, cross-border trade continued and gradually increased due to market forces.

Since cross-border transmission lines must pass through several political jurisdictions, the areas on both sides of the borders and jurisdictions in the line of sight of the transmission projects do not necessarily get any benefits. They face displacement of inhabitants in the line of sight of the proposed transmission line, which naturally creates public and political opposition. This would not only cause delays and cancellation of the projects but also signal risks to future projects. Potential investors would get discouraged. One good example is the Northern Pass transmission project discussed in Section 5 to connect the Canadian province of Québec to the New England electricity grid in the US. The project got canceled after struggling to survive for ten years, losing more than US$300 million in investment. The reason was strong protests from people in New Hampshire who viewed the projects as harming their rural tourism while benefitting Québec (the electricity exporter) and Massachusetts (the importer). This is a common problem for most development projects.

Cross-border electricity interconnection and trade take time. It took more than 100 years in both Europe and North America. In South Asia and Southeast Asia, countries should start cooperating in creating a cross-border electricity market and move forward gradually. Many obstacles and challenges are ahead. However, once it starts, and if there is political will, the process continues. There is no reason to get disappointed if it does not happen in a short period. Stakeholders must make persistent efforts.

6.2 Economics or Market Force Is the Key Driver

The market is the fundamental driver of regional electricity interconnection and trade. If there does not exist a difference in prices between the trading regions, there would be no incentives for electricity trade. The private sector would not be interested in investing in cross-border transmission interconnections if they do not see significant price gaps between the borders to trigger cross-border electricity trade. Governments should not use public investment to build transmission lines connecting the borders if the returns of the project to the society are lower than its opportunity costs. Market force is the main driver of electricity trade between Canada and the US. The north–south electricity trade of Canadian provinces with the US states is higher than the east–west electricity trade between the Canadian provinces themselves, with some exceptions. On the other hand, the electricity trade between the US and Mexico is limited. This is because population and electricity load density, on both sides of the border, are low. If there do not exist markets between South Asia and Southeast Asia, electricity trade may not incur if transmission interconnections are built for political or any reason. See the example of the Middle East and North Africa (MENA) region. The cross-border electricity trade in that region is very small compared to the capacity of the cross-border transmission lines they built (Timilsina & Curiel, 2020).

6.3 No Politicization of Electricity Interconnection and Trade

One of the key lessons to be learned from North America and Europe is to keep the electricity trade away from domestic politics. There are many occasions in which Canadian governments have faced stiff resistance against exporting electricity to the US. The same is true in US states importing electricity from Canada. However, political parties have tried to stay away from messing up the cross-border electricity trade from their national or local politics. The Canadian Constitutions Act of 1867 provides Canadian provinces full authority regarding their electricity business. Therefore, following the market incentives, Canadian provinces have more electricity trade with US states than between themselves. On the US side, too, states and ISOs are free to choose the sources of electricity to meet their demand within the existing rules and regulations. For example, consistent with climate change mitigation goals, California buys clean hydropower from the Canadian province of British Columbia instead of coal-based generation from nearby US states.

Relaxing the central/federal government controls and empowering the power utilities and traders to make decisions on the trade deals is a critical factor for the successful expansion of cross-border or regional electricity trade. Countries in South and Southeast Asia should also give more freedom to their state or provincial governments or electricity utilities regarding their electricity business. If, for example, Indian states have more authority to trade electricity with neighboring countries, there might be a higher level of cross-border electricity interconnections and trade.

A good lesson from Europe is that energy trade must be above national politics. Countries could have border disputes; they may experience conflict with one another; however, energy trade should not be affected by political conflict across borders. Turkey and Greece have a historical conflict regarding Northern Cyprus. Greece and Northern Macedonia have been going through political disputes. Relationships between the Former Yugoslav Republics are not very cordial. However, they have interconnected their electricity transmission systems. They have been trading electricity. This could be a very important lesson for South and Southeast Asia.

6.4 It's More Leadership than Ownership that Matters

The European and North American experiences suggest that both the state-owned electric utilities and the private utilities can equally play roles in investing in cross-border transmission interconnection assets. Public and private utilities also play such a role in engaging in the cross-border electricity trade. Hydro-Québec, BC Power, and Manitoba Hydro, the major exporters of electricity from Canada to the US, are all state-owned utilities. The importers on the other side of the border (in the US) are mostly private utilities. This indicates that incentives, rather than ownership, matters more in electricity interconnections and trade. When there exist business opportunities, entities, irrespective of their ownership type, get engaged in reaping the opportunities. However, leadership and business vision do matter. It's the leadership of these utility companies in Canada and the US, which initiated the cross-border electricity transmission infrastructure and trade.

6.5 Market Reforms Facilitate Cross-Border Electricity Trade

Although we mentioned above that market incentives matter more than utility ownership for cross-border electricity trade, experiences from Europe and North America demonstrate that power sector reforms have significantly facilitated the cross-border electricity trade. The key reason is that market reforms allow electricity traders to quickly grasp market opportunities in a very short interval of time, like a day-ahead market. The fully liberalized market is efficient as it helps to balance supply and demand at the lowest costs. The market reforms harmonize the electricity markets through consistent rules and regulations across the power markets between the borders. In Europe, for example, effective implementation of network code facilitated cross-border interconnection. The TSOs and National Regulatory Agencies (NRAs) have played a key role in avoiding undue discrimination of cross-zonal exchanges and delaying the offering of large cross-zonal capacity after the day-ahead timeframe and being transparent about the capacity calculation methodologies. The day-ahead market coupling and sufficient availability of ID liquidity (e.g., SIDC) are crucial elements in enhancing the efficient use of existing cross-border transmission capacities.

To enhance the market coupling, EU legislation introduced the ITC mechanism, which provides compensation through the ITC fund for the costs of losses incurred by TSOs for hosting cross-border flows of electricity, and the costs of making infrastructure available to host cross-border flows of electricity. The legislation also allows congestion income from day-ahead market coupling to be shared among the capacity owners of the interconnectors if the required interconnection capacities are met and increase the interconnection capacities through network investments (DEA, 2018). At the regulatory level, clear, stable, and non-discriminatory rules for investors in grids and users of the infrastructure are certainly helpful.

6.6 Knowledge and Stakeholder Engagement Is Critical

Another important lesson that can be learned from Europe and North America is that generation and utilization of knowledge and engagement of stakeholders are critical for informed decision-making regarding the cross-border electricity interconnection. Knowledge, generated through analysis and modeling, plays an important role; it helps rationalize the development of cross-border interconnections. It also helps the general public and other stakeholders to understand and realize the benefits they will receive. Both in Europe and in North America, subject experts are always engaged to create and communicate critical knowledge regarding cross-border transmission interconnection initiatives. For example, to make the 2030 target of further integrating the EU electricity markets that specifies the ratio of imported capacity with the total installed capacity of each Member State, the EC set up a Commission Expert Group to provide specific technical advice and to examine any relevant elements that can have an impact on the interconnection target and the development and implementation of interconnectors.

Collaborative decision-making processes are important to build trust and reduce public opposition. Therefore, the development of any new cross-border electricity system must facilitate public involvement, such as citizens, civil society, and relevant stakeholder groups. One key aspect of this is explaining the wider benefits and how this adverse impact minimized the proposed project to all involved communities at the local, national, and regional levels. Since developing a cross-border electricity market involves significant investment and costs, securing the support of all involved parties, based on good cooperation, trust, and simplified procedures, is quite important. Making grants available for studies in the early stages of a project development phase, when the cost–benefit analysis and the technical feasibility of the project are still not clear, is important.

6.7 Regional Institutional Setups

The creation of regional institutions for cross-border electricity integration is needed to develop common rules and regulations to facilitate cross-border electricity interconnection and trade. The EU built several institutions over several decades to move the regional electricity trade agenda forward. Involving electric utilities from various countries, they created the UCPTE in 1951 to facilitate the optimal operation of member countries' power systems. Involving the heads of the national energy regulatory authorities from the EU's 28 Member States, they established the ERGEG in 2003, to assist the EC in consolidating a single EU market for electricity and gas. In North America, the NERC, established in 1968, played a key role in integrating North American electricity markets, particularly those of Canada and the US. Apart from these key institutions, several other agencies and councils have been established over several years, such as ACCER, CEER, and ENTSO-E in the EU, and USCMA, NAAEC in North America, to improve cross-border electricity markets in these regions.

7 Conclusions

The cross-border electricity markets in Europe and North America are almost fully integrated. The European electricity market takes in the interconnected electricity systems of 25 EU Member States and the electricity markets of Britain, Norway, and Switzerland. It is also connected to the electricity markets of neighboring countries, such as Ukraine and Turkey, and the neighboring region, North Africa. In North America, the cross-border electricity markets of Canada and the US are fully interconnected. The electricity systems of the US and Mexico are also interconnected but at a lower scale as compared to US–Canada electricity interconnection. However, it took more than 100 years for both regions to arrive at the current stage of cross-border electricity interconnection and trade. The long experiences of Europe and North America could provide critical lessons for South Asia and Southeast Asia to initiate regional electricity interconnection and trade.

Many studies, observations, and statistics reveal the economics of power sector integration in Europe and North America. The estimates could be several billion US dollars each year. The sources of benefits are many, such as reductions in electricity supply costs through cheaper imports, enhanced markets for suppressed utilization of natural resources used for electricity generation, reductions in costs of meeting climate change targets, avoidance of expensive capacity additions due to the flexibility created by the interconnections, and higher utilization of existing generation capacities due to the interconnections. Some studies have quantified the role of each of these factors in the total benefits. Existing studies also estimated the distributions of economic benefits across the countries involved in regional electricity trade in Europe and North America.

The experiences from Europe and North America suggest that there could be multiple models to invest in transmission interconnections between the borders. It could be regional funding such as those mostly practiced in the EU, or it could be a merchant model where a private sector builds and owns transmission lines connecting two electricity grids, like the Montana–Alberta Tie Line in North America. State-owned electricity utilities in Canada, such as Hydro-Québec, are major investors of cross-border electricity transmission lines between Canada and the US.

While there are many drivers of cross-border electricity trade both in North America and Europe, such as the reduction of electricity supply costs to meet demand, the demand for clean power, particularly hydropower, to meet the climate change mitigation targets is becoming the primary factor for the enhanced electricity trade more recently.

Although both Europe and North America are in the most advanced stage in terms of electricity market interconnection, they still face many challenges. Investors face a long lead time to prepare and develop cross-border transmission interconnection projects as they must go through a long clearance process from the local, provincial, and federal government authorities. Often, they receive opposition from the inhabitants who come across the line of sight of the proposed transmission lines. The problem gets amplified in jurisdictions where the transmission systems do not serve instead just pass through to serve other jurisdictions. Project delays and cancellations discourage private investors. Electricity transmission projects are capital-intensive infrastructure projects. Governments' budget allocations would not be sufficient for the projects. Private investors perceive high risks due to uncertainties and delays due to long clearance processes, land acquisition challenges, and potential opposition by the population in the line of sight. Therefore, cross-border transmission lines have not been built to satisfy the demand both in Europe and North America. The complex regulatory regime and market rules are not favorable for new investors in doing business in cross-border electricity interconnection and trade.

Many valuable lessons can be learned from the experiences of Europe and North America. Regional electricity interconnection and trade take decades; it goes through a series of ups and downs. However, persistence efforts ultimately bring success. It is the market forces than anything else that drives the cross-border electricity trade. However, political will and cooperation are necessary to realize the benefits that can be accrued through cross-border electricity trade. Without political leadership and the participation of other stakeholders, particularly the private sectors and electricity utilities, the development of cross-border transmission interconnections is not possible. The North American and European experiences suggest that providing more authority to state and local governments is instrumental in developing regional electricity interconnection and trade. Power sector reforms played an important role in both regions to integrate the electricity markets across the borders. Experience from Europe and North America also suggests that the creation of knowledge, electricity price reforms, and regional institutions are the initial steps to initiate regional electricity interconnections and trade.

Notes

1 The EU electricity market includes 27 EU Member States, the UK, Norway, and Switzerland.
2 Under the Paris Climate Agreement, various countries set targets to reduce, voluntarily, their GHG emission by 2030. The EU's target is 40% below 1990 level; the US's target is 50% to 52% below 2005 level, Canada's target is 30% below 2005 level; and Mexico's target is 25% below 2013 level (UNFCCC, 2020).
3 The deviations between scheduled flows and physical flows as unscheduled flows. Loop flows are unscheduled flows stemming from scheduled flows within a neighboring bidding zone, whereas transit flows are unscheduled flows stemming from a scheduled flow between two or more bidding zones or control areas (THEMA, 2013).
4 The EU Regulation on Wholesale Energy Market Integrity and Transparency (REMIT) is designed to increase the transparency and stability of the European energy markets while combating insider trading and market manipulation.

References

Abrell, J., & Rausch, S. (2016). Cross-country electricity trade, renewable energy, and European transmission infrastructure policy. *Journal of Environmental Economics and Management*, 79, 87–113. https://doi.org/10.1016/j.jeem.2016.04.001.
ACCER and CEER. (2019). Annual Report on the Results of Monitoring the Internal Electricity and Natural Gas Markets in 2018 Electricity Wholesale Markets Volume. https://acer.europa.eu/Official_documents/Acts_of_the_Agency/Publication/ACER%20Market%20Monitoring%20Report%202018%20-%20Electricity%20Wholesale%20Markets%20Volume.pdf
ACER. (2019). *Annual activity report 2018*. ACER: Ljublijana.
ACER and CEER. (2019). *Annual report on the results of monitoring the internal electricity and natural gas markets in 2018: Electricity and gas retail markets volume*. Brussels: Agency for the Cooperation of Energy Regulators (ACER): Ljubljana and Council of European Energy Regulators (CEER).

Antweiler, W. (2016). Cross-border trade in electricity. *Journal of International Economics*, *101*, 42–51. https://doi.org/10.1016/j.jinteco.2016.03.007

Booz & Company, Newbery, D., Strbac, G., Pudjianto, D., Noel, P., & Fisher, L. (2013). *Benefits of an integrated European energy market*, Final report. Amsterdam: Booz and Company: .

CEER. (2020). *Annual report 2019*. Brussels: CEER.

Chen, Y.-K., Koduvere, H., Gunkel, P. A., Kirkerud, J. G., Skytte, K., Ravn, H., & Bolkesjø, T. F. (2020). The role of cross-border power transmission in a renewable-rich power system – A model analysis for Northwestern Europe. *Journal of Environmental Management*, *261*, 110194. https://doi.org/10.1016/j.jenvman.2020.110194.

CRS. (2017). *Cross-border energy trade in North America: Present and potential. R44747*. Washington, DC: Congressional Research Service (CRS).

DEA. (2018). *European experiences on power markets facilitating efficient integration of renewable energy*. Copenhagen: Danish Energy Agency (DEA).

DOE. (2016). *Electricity in North America: Baseline and literature review*. Washington, DC: The US Department of Energy (DOE).

DOE. (2017). *Enhancing electricity integration in North America*. Washington, DC: DOE.

EC. (2017). Towards a sustainable and integrated Europe. Report of the commission Expert Group on Electricity Interconnection Targets. Brussels: European Commission (EC).

EC. (2019). *Investing in European networks: The connecting Europe facility -- Five years Supporting European infrastructure*. Brussels: EC.

E.DSO. (2020). DSO services for our customers: "lead the transition – serve the customers". E.DSO white paper. Brussels: European Distribution System Operators (E.DSO).

EIA. (2020). *Wholesale electricity and natural gas market data*. Washington, DC: Energy Information Agency (EIA).

ENTSO-E. (2019a). *ENTSO-E annual report 2018*. Brussels: European Network of Transmission System Operators for Electricity (ENTSO-E).

ENTSO-E. (2019b). *Statistical factsheet 2018*. Brussels: ENTSO-E.

GAO. (2018). *North American energy information. Information about cooperation with Canada and Mexico and among U.S. agencies*. US Government Accountability Office (GAO): Washington, DC.

GTM. (2017). *Enhancing grid connectivity in North America: Expanding power trade*. New Delhi: Global Transmission Report (GTM).

Huang, D., & Van Hertem, D. (2018). Cross-border electricity transmission network investment: Perspective and risk framework of third party investors. *Energies*, *11*(9), 2376. https://doi.org/10.3390/en11092376.

IEA. (2019). *Integrating power systems across borders*. Paris: International Energy Agency (IEA).

IEA. (2020). *Data and statistics*. Paris: IEA.

Jamasb, T., & Pollitt, M. (2005). Electricity market reform in the European Union: Review of progress toward liberalization & integration. Center for Energy and Environmental Policy Research Working Paper No. 05-003. Massachusetts Institute of Technology (MIT).

Krupnick, A., Shawhan, D., & Hayes, K. (2016). Harmonizing the electricity sectors across North America. Recommendations and action items from Two RFF/US Department of Energy Workshops, Discussion Paper. Washington, DC: Resources for the Future (RFF).

Manitoba Hydro. (2013). *The Manitoba hydro system: Interconnections and export markets.* Winnipeg.

Martin-Nielsen, J. (2009). South over the wires: Hydro-electricity exports from Canada, 1900–1925. *Water History, 1*(2), 109–129. https://doi.org/10.1007/s12685-009-0011-6

NEB. (2020). *Online database.* Calgary: National Energy Board (NEB).

NERA. (2020). *Mexican electricity wholesale market report 2019.* New York: NERA Economic Consulting.

NERC. (2020). *NERC 2019* Annual Report. Washington, DC: North American Electric Reliability Corporation (NERC).

Newbery, D., Strbac, G., & Viehoff, I. (2016). The benefits of integrating European electricity markets. *Energy Policy, 94,* 253–263. https://doi.org/10.1016/j.enpol.2016.03.047

Rumpf, J. (2020). Congestion displacement in European electricity transmission systems – Finally getting a grip on it? Revised safeguards in the Clean Energy Package and the European network codes. *Journal of Energy and Natural Resources Law, 34*(4), 409–436. https://doi.org/10.1080/02646811.2019.1707441

Steinbacher, K., Schult, H., Jörling, K., Fichter, T., Staschus, K., Schröder, J., & Lenkowski, A. (2019). *Cross-border cooperation for interconnections and electricity trade: Experiences and outlook from the European Union and the GCC.* Berlin: Navigant Energy Germany.

THEMA. (2013). *Loop Flows – Final Advice.* Report Prepared for the EC. Brussel: THEMA.

Timilsina, G. R., & Curiel, I. F. D. (2020). *Power system implications of subsidy removal, regional electricity trade, and carbon constraints in MENA economies.* World Bank Policy Research Working Paper, WPS 9297. Washington, DC: World Bank.

UCTE. (2003). *The 50 year success story – Evolution of a European interconnected grid.* Brussels: Union for the Coordination of Transmission of Electricity (UCTE).

UN. (2020). *UN comtrade database.* New York: United Nations (UN).

UNFCCC. (2020). *NDC registry.* Bonn: United Nations Framework Convention on Climate Change (UNFCCC).

WPC. (2019). *Annual report.* Ottawa: Water power Canada (WPC).

Zachmann, G. (2013). *Electricity without borders: A plan to make the internal market work. Bruegel blueprint Series 20.* Brussels: Bruegel.

8

POWER MARKET INTEGRATION IN AUSTRALIA

Attributes and Frameworks for Sustainable Cross-Border Cooperation

Adil Khan Miankhel and Kaliappa Kalirajan

1 Introduction

Australia, with its six states and two territories, has been characterized as having post-war protectionist policies with wage fixation and subsidies that influenced almost all sectors of the highly regulated economy prior to the 1980s (KPMG, 2013). The move away from a highly regulated economy started with the stepwise removal of tariffs and quotas and with the free floating of the Australian dollar in 1983. Consequently, these reforms exposed the tradable sector to more international competition. The recognition of the need to reform the economy to make it competitive directed focus to the non-tradable sectors of the economy too. Hence, between the late 1980s and early 1990s some initial reform measures were undertaken in electricity, rail, road, telecommunications, and water.

In the context of the power sector in Australia, prior to the 1990s, the state governments that supplied electricity to local governments and private businesses owned vertically integrated utilities. It was estimated that in terms of direct business cost, electricity contributed 1% to business costs overall, excluding electricity-intensive industries such as aluminum smelting, and it outweighed telecommunications, post, gas, transport, and water costs. When indirect costs of electricity were also taken into account, the total influence of electricity on business costs amounted to 3–5% (Siemon, 1995). Besides, the state-run vertically integrated utilities were operating inefficiently. For example, in 1990, Victoria's State Electricity Commission (SEC) employed 20,000 employees to cater to a maximum demand of about 6,000 megawatts (MW). Meanwhile, customers went off the supply for an average of 510 minutes compared to 140 minutes today (Skinner, 2018). The

DOI: 10.4324/9781003433163-8

Victorian government assessed its performance to be unsatisfactory as its electricity charges were too high and it had too much debt. Overall, the government concluded that the operation of the SEC was inefficient (Stockdale, 1994). The SEC was not alone in receiving this assessment: most of the state utilities were seen as inefficient and largely unnecessary investments, which were not guided by the prices and were the result of poor planning (Skinner, 2018). The risks associated with the inefficiency of the sector were borne by the government and customers (Pierce, 2013).

Victoria initiated competition in the energy sector. Stockdale (1994) emphasized that the Victorian Government wanted to keep energy prices down as a way of attracting investment and maintaining its low-cost energy advantage. Siemon (1995) argued that such a process of industrialization required more micro-economic reforms. The objective of those reforms would be to unleash the forces of competition to ensure lower energy prices for all sectors. Those micro-economic reforms were also expected to improve productivity, employment and the living standards of the workforce. While analyzing the issue of debt, Siemon (1995) indicated that even though debt reduction was a lower priority for the government, the debt to equity ratio was declining and was achieved by expanding sales and incomes rather than by repayment of principal.

The following section provides a synopsis of the evolution of the Australian National Energy Market governance structure, including the reform measures undertaken over the years. The integration models of the power market in Australia will be explained in the following section. The next section describes the functioning of the National Electricity Market (NEM) of Australia. The evolution of the investment models in the energy sector is then analyzed in the following section. A final section brings out the lessons that the North East Region (NER) can learn from the Australian experience of the process of power market integration.

2 National Electricity Market of Australia

The National Electricity Market (NEM) is one of the world's longest integrated power systems, spanning about 5,000 kilometers and connecting electricity markets in the southern and eastern coastal regions of Australia. The NEM is composed of the five interconnected states of Queensland (QLD), New South Wales (NSW) (which includes the Australian Capital Territory (ACT)), South Australia (SA), Victoria (VIC), and Tasmania (TAS), which also act as five pricing regions. There are over 300 registered participants in the NEM consisting of generators, transmitters, and distributers. Table 8.1 gives an overall picture of the structure of the NEM.

The grid demand (including scheduled and semi-scheduled generation, and intermittent wind and large-scale solar generation) was at its peak

TABLE 8.1 National electricity market at a glance

Participating jurisdictions	QLD, NSW, VIC, SA, TAS, ACT
NEM regions	QLD, NSW, VIC, SA, TAS
NEM installed capacity (including rooftop solar PV)	60 839 MW
Number of large generating units	250
NEM turnover 2018–2019	$19.4 billion
Total electricity demand 2018–2019	205 TWh
National maximum demand 2018–2019	33 941 MW

Source: AER; AEMO; Energy Made Easy; Victorian Essential Services Commission; CER (state of Energy market (2019)

TABLE 8.2 NEM grid demand (terawatts hours)

Year	Grid Demand	Rooftop Solar
2005–2006	202.76	0.01
2006–2007	207.04	0.01
2007–2008	208.62	0.02
2008–2009	210.48	0.04
2009–2010	209.84	0.15
2010–2011	207.49	0.69
2011–2012	203.38	1.73
2012–2013	198.21	2.96
2013–2014	193.61	4.02
2014–2015	193.97	5.05
2015–2016	197.60	6.10
2016–2017	196.50	6.30
2017–2018	196.16	7.30
2018–2019	195.69	9.17

Source: State of Energy Market (2019)

(210.48 Terawatts hours) during 2008–2009 (Table 8.2). The grid demand has been decreasing with the introduction of rooftop solar, which increased during 2017–2019. While Table 8.3 shows the total generation capacity of NEM by fuel source at the national level, Table 8.4 shows the statewise total generation capacity by fuel source in NEM.

The share of wind and solar in the NEM generation has been increasing over time. The overall share of wind has increased from 0.2% to 7.6%, while rooftop solar has increased from 0.1% in 2009–2010 to 4.5% in 2018–2019. Table 8.5 shows that there has been a considerable amount of capacity investments over the recent years in wind, solar and battery storage.

The separation of wholesale, network and retail businesses in the 1990s led to the integration of many retailers and generators to become 'gentailers'

TABLE 8.3 Generation capacity of NEM by fuel source

	Output		Capacity	
	GWh	% of Total Output	MW	% of Total Capacity
Black coal	109,940	54	18,346	30
Brown coal	34,589	17	4,660	8
Gas	15,813	8	12,065	20
Hydro	15,261	7	7,983	13
Wind	15,638	8	6,141	10
Solar farms	3,419	2	2,762	5
Other dispatched	1,039	1	999	2
Rooftop solar	9,168	4	7,883	13

Source: State of Energy Market (2019)
Note: Other dispatched includes battery, biomass, waste gas, and liquid fuels. Storage includes only battery storage.

TABLE 8.4 Statewise total generation (%) of NEM by fuel sources

	Black Coal	Brown Coal	Gas	Hydro	Wind	Solar Farm	Rooftop Solar	Battery Storage	Other Dispatched
Queensland	48%	0%	22%	4%	1%	8%	15%	0%	2%
NSW	52%	0%	12%	14%	8%	4%	11%	0%	1%
Victoria	0%	34%	18%	17%	15%	3%	13%	0%	0%
South Aus.	0%	0%	46%	0%	29%	4%	16%	2%	3%
Tasmania	0%	0%	12%	73%	10%	0%	5%	0%	0%

Source: State of Energy Market (2019)
Note: Other dispatch includes battery, biomass, waste gas, and liquid fuels. Storage includes only battery storage.

TABLE 8.5 Capacity (MW) investments by sources

	Black Coal	Brown Coal	Gas	Hydro	Wind	Solar	Battery Storage
2012–2013	60				588		
2013–2014					567		
2014–2015	–1174	–189			327	102	
2015–2016	–190	–920	–18			109	
2016–2017	–80	–1600			464		
2017–2018				70	978	465	100
2018–2019	0	0	0	0	1,540	2,401	90

Source: State of Energy Market (2019)

to minimize the need to hedge their positions in futures (derivatives) markets. However, such a business strategy may drain liquidity from derivatives markets, which would affect the entry and expansion of businesses that were not vertically integrated. AGL Energy, Origin Energy, and Energy Australia are the major businesses, which collectively owned or controlled about 70% of new electricity generation between 2011 and 2018. These businesses also have interests in upstream gas production and storage, which has complemented their stakes in gas-fired power plants and energy retailing.

AGL Energy, Origin Energy, and Energy Australia established retail arms as well as generation. The businesses include Engie (which established Simply Energy), Alinta, ERM Power, Meridian Energy (Powershop), and Pacific Hydro (Tango). These businesses supplied electricity to over 66% of small customers and 77% of small gas customers in southern and eastern Australia in June 2018 (Australian Energy Regulator, 2018). Government-owned generators are also vertically integrated. While the Commonwealth of Australia (through Snowy Hydro) owns the retailers – Red Energy and Lumo Energy, the Tasmanian Government owns Momentum Energy through Hydro Tasmania. The largest standalone electricity retailers operating in the NEM are Amaysim (trading under its own name and as Click Energy) and M2 Energy (trading as Dodo Power and Gas, and Commander Power and Gas) respectively, with 1.4 and 1.1% of small customers across the NEM (Australian Energy Regulator, 2019).

There are limitations in the physical transfer capacity in the NEM. For example, while NSW has connections for export to and import from VIC, VIC can import from and export to NSW, SA and TAS; QLD, with two interconnectors, has connections only with NSW; and TAS can import from and export to VIC. There is currently no direct connector between NSW and SA. The interconnectors become congested when there is peak demand, and the NEM is separated into its regions. Such a separation would promote price differences across markets and exacerbate reliability problems for regional utilities (ACCC, 2000, IEA, 2001). Due to these constraints, the regional spot markets are somewhat less than perfectly integrated. Generally, electricity is sold from a lower price region to a higher price region when there is full capacity operation of interconnectors. However, consumer prices are not equalized across regions. Importing electricity from interconnectors serves as a shock absorber for the local regions, which cannot increase the capital stock for generation within a short period of time (AEMC, 2020).

Generators and retailers trade electricity in the NEM through a gross pool managed by the Australian Energy Market Operator (AEMO). Apart from keeping generating capacity in reserve, the AEMO matches electricity production with electricity consumption, keeping in mind transmission limitations, to determine the energy price. In delivering electricity, a dispatch price is determined every five minutes. The spot price is determined by averaging six dispatch prices every half hour for each NEM region. The AEMO uses the spot price as its basis for settling the financial transactions for all electricity traded in the NEM (AEMC, 2020).

If there are system limitations or increases in consumption, the AEMO makes adjustments. For example, when supplies are inadequate to meet consumption, the AEMO may issue notices to the market for additional generation or directly intervene as a last resort. The AEMO also monitors electricity voltage and frequency to make sure the system remains secure, the AEMO also regularly maintains surveillance over the impact of planned power outages to ensure the system can accommodate any loss of generation or transmission capacity (Higgs et al., 2016).

The AEMO uses the spot market to match the supply of electricity from power stations with real-time consumption by households and businesses. When the spot price shows an increase, generators would increase their output or more expensive generators would turn on to sell extra power to the market. Spot prices, which vary across regions, are expected to be updated every five minutes in 2021 (AEMC, 2020). The spot price for wholesale energy often fluctuates and it could range between minus $1,000 per MWh to $14,500 per MWh. Retailers enter into hedging contracts to minimize their financial risks related to wholesale energy costs, which could be for a year or for several years. On the other hand, generators would stabilize their revenue stream through hedging contracts. Such hedging contracts would facilitate them to improve their credit worthiness for securing financing from financial institutions for long-term investments. Generators will be liable to financial penalty when they are unable to provide electricity to the retailers as per the agreed contract price. The higher the spot price, the higher the financial penalty. Due to this aspect, generators are forced to hold a certain amount of energy in reserve to serve the retailers when the demand is high. This process establishes a link between financial and physical markets. Nevertheless, this link is related only to financial derivative contracts (swaps and caps).

This is not the case in Power Purchase Agreements (PPAs), which are mostly used by wind and solar farms. These farms usually sell renewable energy certificates to the purchasers at a fixed price. In PPAs, there is no incentive for sellers to manage their generation based on spot price fluctuations in the market. However, PPAs incentivize sellers to produce as much electricity as possible. Financial contracts, of which swaps and caps are core types, are traded either on the Australian Stock Exchange or bilaterally. Table 8.6 presents the trade volumes of electricity future contracts in the NEM.

The National Electricity Rules (the Rules) set a maximum spot price, also known as the Market Price Cap (MPC), which would be adjusted annually

TABLE 8.6 Traded volumes in electricity futures contracts

Volume (TWh)

	2009–2010	2010–2011	2011–2012	2012–2013	2013–2014	2014–2015	2015–2016	2016–2017	2017–2018	2018–2019
Futures	399	549	437	342	387	446	396	398	338	481
OTC contracts	221	315	227	291	251	73	111	119	109	
NEM generation volume	210	208	203	198	194	194	198	196	196	196

Futures as % of NEM output

	2009–2010	2010–2011	2011–2012	2012–2013	2013–2014	2014–2015	2015–2016	2016–2017	2017–2018	2018–2019
Futures	190	264	215	173	199	230	200	202	173	246
OTC contracts	105	151	112	147	129	38	56	60	56	

Source: State of Energy Market (2019)

TABLE 8.7 Annual NEM electricity prices

| | Volume Weighted Average Spot Prices ($/MWh) | | | | |
	Queensland	NSW	Victoria	South Australia	Tasmania
2000–2001	44.52	40.69	48.56	66.55	
2001–2002	37.63	38.32	32.77	33.5	
2002–2003	41.35	37.07	29.62	32.74	
2003–2004	30.74	36.71	26.81	39.29	
2004–2005	30.52	45.51	28.78	39.25	
2005–2006	31.4	43.08	36.07	43.91	59.32
2006–2007	57.17	67.22	60.57	58.72	50.86
2007–2008	57.91	44.45	50.61	101.29	56.54
2008–2009	36.3	42.61	49.11	68.54	61.9
2009–2010	37.39	52.27	42.15	82.5	30.36
2010–2011	33.89	42.98	29.11	41.94	31.06
2011–2012	30.05	30.52	28.27	32.03	32.91
2012–2013	70.3	56.04	60.79	74.39	48.75
2013–2014	60.54	53.15	54.24	68.25	42.19
2014–2015	61.28	36.28	31.65	42.31	37.29
2015–2016	64.1	54.4	50.11	67.07	97.14
2016–2017	102.65	88.21	69.5	123.15	75.88
2017–2018	74.61	84.59	99.26	108.72	87.57
2018–2019	83.06	92.38	123.73	128.28	88.36

Source: State of Energy Market (2019)

for inflation. The Rules also set a minimum spot price, called the market floor price. The Australian Energy Market Commission's Reliability Panel has the authority to review the market price cap and market floor price every four years to ensure they are in line with the NEM reliability standard (AEMC, 2020). Table 8.7 given below states the annual NEM electricity prices by region.

The ageing coal-fired power generation is expected to be replaced over the coming years in Australia. AEMO in its Integrated System Plan forecasts its transmission requirements to connect new generators to ensure a reliable supply of power to the consumers over the next 20 years. Transmission businesses, drawing on their local knowledge, choose the best project to meet the network's need identified in the Integrated System Plan. Australian Energy Regulator (AER) approves the Regulatory Investment Test for Transmission (RIT-T) and sets network prices, so customers are only paying for investment, which is efficient. The AEMO also publishes the following reports: (i) National Electricity Forecast Report, which is annual energy and maximum demand forecasts over the next ten years by the NEM region; (ii) Electricity Statement of Opportunities, which assesses supply adequacy in the NEM over the next ten years; (iii) reports highlighting opportunities for generation and demand-side investment; and (iv) the constraint reports detailing interconnector capacity and constraints in the transmission network (AEMC, 2020).

Each region of the national electricity market has a different jurisdictional transmission planning body. NSW and ACT have TransGrid; Queensland has Powerlink; South Australia has ElectraNet; Tasmaniaihas Transend; and Victoria has the AEMO (in its role as Victorian transmission network service provider). Transmission businesses must plan and develop their networks in line with reliability standards, particularly to minimize costs. However, these businesses can undertake investments, where those investments would result in a net market benefit but are not necessarily designed to meet a specific reliability standard.

Similarly, the national distribution framework requires conducting a distribution annual planning review; distribution annual planning report (DAPR); and demand-side engagement obligations. It also conducts distribution investment project assessment in the form of a regulatory investment test for distribution (RIT-D) and dispute resolution. Currently, there are 18 distribution network service providers (DNSPs) concerning electricity and 11 are required to undertake an annual planning process covering a minimum forward planning period of five years for its distribution assets (and ten years for dual function assets).

3 The Evolution of the Australian Energy Market Integration

Power generation requires substantial capital investment in power stations over a long period before power can be sold. Different power stations require different lengths of time to become operational. For example, power stations using brown coal power plants require the most time, and gas turbine-driven stations require relatively little time for operation. These investments at times may not yield the expected profits (Siemon, 1995). An example is that of the construction of the Loy Yang project in Victoria in the 1970s and 1980s. The construction was done to double the SEC generation capacity, which proved to be a mistake as demand failed to meet the original projections. The SEC was not alone in this phenomenon, as in other industrialized countries lower growth in demand for electricity also caught utilities, both publicly and privately owned, on the wrong foot (Fitzgerald, 1995).

As almost all the Australian states exhibited similar inefficient governance issues, such as an overstaffed and highly unionized labor force, undertaking reforms across different states was highly challenging. Against this backdrop, the governments decided to embark on a reform process by introducing competition in the supply of electricity to industrial and commercial customers in 1990 by following the examples of England and Wales instead of the US preference for separate energy and capacity markets. New South

Wales, Victoria, and South Australia were interconnected; however, these links were only opportunistically used to reduce fuel costs when it suited both parties to reduce costs and were acting independently (Skinner, 2018).

In May 1990, the Commonwealth requested the Industry Commission (now Productivity Commission) to inquire into the generation, transmission and distribution of electricity and the transmission and distribution of gas. The Industry Commission in its Energy Generation and Distribution report (1991) found that the electricity and natural gas sectors were not performing at their optimal level due to poor investment decisions of the 1970s and 1980s that resulted in overcapacity in generation and overstaffing of utilities. Therefore, electricity and gas were not supplied at the least cost. The report was tabled to the Council of Australian Governments (COAG). Drawing on this report, the COAG established a National Grid Management Council (NGMC) to coordinate the planning, operation and development of a competitive electricity market. In this way, the formal process of developing the National Electricity Market began in 1991. Moreover, the major tax reforms pursued during that period resulted in yielding a benefit of 0.5% of GDP compared to energy sector reforms that were likely to provide a benefit of 1.5% of GDP. Therefore, micro-economic reforms in the energy sector provided a major incentive to states to pursue them (KPMG, 2013). These reforms mainly concerned the restructuring of the electricity supply industry with the vertical separation of generation and retail from the natural monopoly elements of transmission and distribution. Attempts were made to improve efficiency through the privatization of publicly owned electricity generation and electricity transmission and distribution assets. The unification of the interconnected systems of New South Wales, ACT, Victoria, and South Australia with the systems of Queensland and Tasmania was also one of the objectives of the reforms (KPMG, 2013).

The release of this report focusing on efficiency gain through micro-economic reforms also necessitated the need for a national competition policy. Because of a later Inquiry on National Competition Policy that resulted in a report in 1993, the COAG endorsed the Hilmer competition policy reforms in 1995. There was a smooth transition toward the reform process with NSW and Victoria taking the lead. In 1991, the Electricity Commission of NSW was renamed Pacific Power, resulting in six units comprising three generating groups, a pool trading unit, a transmission network business unit, and a services unit. Pacific Power also established an internal power market, namely ELEX. Similarly, the SEC of Victoria also established VicPool II in 1992 to trial an internal power market. During this micro-economic reform process, the bottom-up reform approach was adopted to minimize the risks before the functioning of NEM at a broader scale. Accordingly, states initiated the process with local trials of competitive markets, establishing an independent economic regulator and supporting the legal and regulatory changes.

Businesses also actively participated in the reform process by providing technical and economic advice to the government. The government also involved all stakeholders by ensuring transparency in the whole reform process in the form of workshops and seminars. Each state analyzed the impact of the competition reforms in electricity (e.g. Victoria Status Report 1993, NSW 1995 Electricity Reform Statement, QLD Electricity Industry Structure Task Force Report December 1996, and the Industry Commission Report for South Australia 1996) and tailored the design to its specific requirements excluding the wholesale market trading design (KPMG, 2013).

The NGMC produced a National Grid Protocol (NGP) establishing a set of rules, responsibilities and technical requirements for connecting to the national grid and participating in the trade of bulk electricity. This was initially limited to generators and large customers. The NGP adopted a cautious approach to minimize the transitioning risk. The NGMC, with the agreement of the COAG, conducted a Paper Trial simulation of a national electricity market from November 1993 to June 1994 involving a large number of participants (approximately 170) to assess the operation of an electricity market in the Australian context. This exercise benefited the non-experimenting regional markets' jurisdictions, namely Queensland, ACT, South Australia, and Tasmania. This process highlighted a number of important issues such as emphasizing the development of commercial and reliable information systems before the commencement of the market.

In 1994, the NGMC set up the Market Steering Committee and various NGMC working groups.[1] The National Electricity Code (Code) Working Group (WG), in consultation with other WGs, outlined market governance arrangements with market trading rules, power system security rules, third party access arrangements for transmission and distribution networks, and metering rules. It took a long time for the NGMC to release the first version of the National Electricity Code for public comments. The first version of the Code was released in late 1995, and was later called the NEM rules. In 1995/1996, the NGMC developed the institutional and governance arrangements for the market. First, it established the National Electricity Market Management Company (NEMMCO) as the market and system operator. This was followed by the establishment of the National Electricity Code Administrator (NECA) to manage the Code change process, monitor Code compliance and enforce breaches through a National Electricity Tribunal. The responsibility of the NEMMCO was to not only operate the physical dispatch process across the NEM and register Code participants, but also to plan and coordinate for power system security. The process gradually transferred the control of the electricity businesses away from the state governments. Such a transfer was considered essential for attracting private investment (KPMG, 2013).

Harmonization of laws and regulations was the next necessary step, as the Commonwealth had no constitutional authority over electricity. In order to seek legal authority, the Commonwealth required agreement from the participating jurisdictions. The agreement was achieved through COAG, which endorsed a cooperative legislative scheme with South Australia being the lead legislator for a National Electricity Law (NEL). Each participating jurisdiction agreed to adopt legislation identical to that of the lead legislature (i.e. South Australia) and not to change or repeal the cooperative legislation without unanimous consent. The Australian Competition and Consumer Commission (ACCC) performed transmission regulation, while state regulators performed Distribution and Retail regulation. This whole structure was empowered by the NEL. As a result, the NEL was legislated in South Australia and was reflected in each participating jurisdiction's Electricity Act (Skinner, 2018).

Finally, on 13 December 1998, the NEMMCO took over the controls of the power system in Queensland, New South Wales, Victoria, South Australia, and the Australian Capital Territory; and the National Electricity Code came into force. Tasmania participated as a separate trader. In 2001, COAG established a new Ministerial Council on Energy (MCE) comprising federal, state and territory energy ministers to provide a forum for national leadership on energy issues. In 2004, COAG approved the Australian Energy Market Agreement establishing the new national governance, regulatory and legislative framework of the Australian Energy Market (and the NEM) (KPMG, 2013). Tasmania joined the NEM in 2005.

In the subsequent years, a number of governance changes took effect. In 2005, the Code became the NEM Rules, subordinate to the NEL. The NECA was abolished and the Australian Energy Market Commission (AEMC) was created to manage them. In 2009, NEMMCO merged with several gas market operators to become the Australian Energy Market Operator. In 2011, COAG merged the MCE with the Ministerial Council on Minerals and Petroleum Resources (MCMPR) into one body called the Standing Committee on Energy and Resources (SCER). In 2017, the Energy Security Board (ESB) was created to coordinate the other institutions and recommend expedited changes to governments.

With the above governance changes, the market objectives in the Code were replaced by a single National Electricity Objective (NEO) in the NEL. The idea behind the NEL objective is that if the NEM is efficient in an economic sense, the consumers' long-term economic interests in respect of price, quality, reliability, safety, and security of electricity services will be maximized (AEMC, 2020). The new NEL enables any person, including industry participants and end users, to initiate a Rule change proposal. The

AEMC can initiate any changes to the Rules only to correct a minor error or of non-material kind. Nevertheless, on the other hand, its role is to manage the process of Rule change and to consult to decide on Rule changes proposed by others. Moreover, regarding the market development function, the AEMC must conduct reviews into any matter related to the national electricity market. This strict policy control on market development was introduced because of the importance of attracting finance to the Australian energy sector at competitive rates.

The National Energy Retail Rules (NERR) provide the governance structure for the sale and supply of energy (electricity and natural gas) from retailers and distributors to customers. The Rules have been made by the AEMC under the National Energy Retail Law and while some have a general application, most of the rules address the sale and supply of energy to residential and other small customers, including the key electricity consumer protection measures and model contract terms and conditions. While formulating the Rules, the AEMC ensures that the proposed changes meet the National Energy Retail Objective and protect small consumers. Besides providing consumer protection and model contracts, the Rules ensure facilitating customer connections and retail competition by allowing customers to choose between competing retailers and to switch their retailer. Moreover, it also facilitates in terms of energy-specific consumer protections and basic terms and conditions contained in standard and market retail contracts (AEMC, 2020).

4 The Integration Models of the Power Market in Australia

There are different drivers for integration of cross-border power systems and the primary driver is economics; that is, to reap the benefits of scale economies and reduce the overall investment and operating costs of the power systems. This brings in the question of how to ensure the reliability and security of the power systems in the light of evolving environmental challenges. The integration of the cross-border power system facilitates maximization of benefits, such as sustaining security benefits and integration of increasing shares of variable renewable energy (VRE) sources. For example, the security benefits are obtained due to larger systems having a diverse mix of supply and demand resources, which are shared among the participants. Moreover, larger power systems are able to integrate higher shares of variable resources and with larger balancing areas, the smoothing of the underlying resource could be better achieved. Through interconnectors across borders, the flexibility of resources in achieving the objectives of integration of power systems could be enhanced.

There could be a number of challenges, as the integration process would produce winners and losers. Therefore, the challenge could be how to allocate the benefits across participants when differential investment (cross-border transmission infrastructure or local infrastructure investments that have cross-border implications) and operational costs are involved across the jurisdictions. Regarding security challenges, the first is that each jurisdiction needs to be self-sufficient. The second is that the tight coupling of power system across borders increases the risks of blackouts and make the integrated system more vulnerable. The third is that with the integration of variable renewables the integrated systems could be exposed to unexpected cross-border power flows ('loop' or 'transit' flows). Moreover, harmonization of policies also poses a challenge as support to increase local investments in variable renewables could also increase the uncoordinated cross-border power flows. The local capacity investments might also result in overcapacity in one jurisdiction relative to total system requirements (IEA, 2019). Increased coordination in dispatch not only increases the energy smoothening objectives of least investment and operating costs, but also increases the security of integrated power systems. The lack of such coordination could result in misalignment in investments with underinvestment in one region and more in other region to limit cross-border flows. This also requires developing a coordinated reliability framework.

In addition to system operations and governance, resource adequacy is also an important component of cross-border integration to ensure sufficient long-term investment in transmission and generation. The long-term investments also necessitate regional planning and agreements on investment cost-sharing. This brings in an important aspect of regional resource adequacy and to achieve it, thei'beneficiary pays' principle may be followed in which costs are shared in proportion to each party's received benefits. In the cross-border integration process, it is important to agree on how to measure and allocate interconnector capacity once interconnectors are in place. Moreover, interconnector capacity should be allocated across multiple periods to ensure reliable access and achieve the optimal utilization of interconnector capacity in real time. Effective allocation of interconnector capacity is important for jurisdictions wishing to engage in capacity products as opposed to energy products trade across borders (IEA, 2019).

There are two ways to look at the cross-border integration process of power systems. One dimension extends from limited (bilateral, unidirectional power trades) to complete (unified market and operations) integration; the greater the degree of integration, the greater the potential benefits. However, more integration also involves the complexity of organizations. On this spectrum, the models of integration can be categorized into three major groups namely: bilateral, multilateral, and unified. In bilateral integration, exchange of power takes place between only two jurisdictions, and in a few instances, this trade may be unidirectional; in others, it may involve an intermediary transit (or wheeling) jurisdiction that facilitates transfer flows of power, but is not party to the transaction. In the multilateral integration mode, three or more jurisdictions are involved in the power trade among themselves. The process of integration is supported through local regional institutions that coordinate or manage the flow of power trade; however, these institutions do not replace the local institutions. The market structures can vary within each jurisdiction. In the unified model of integration, regional institutions assume some or all of the functions for managing the power system across multiple jurisdictions, including as a minimum market organization, and maybe even system operations (IEA, 2019).

The second way of looking at the cross-border integration concerns time-based dimension. The time-based dimension may involve long-term planning or power purchase agreement to short-term arrangements, such as ancillary services and real-time dispatch. Between these two extreme points in the time-based spectrum, there could be arrangements in the form of sharing short-term forecasts or information on day-ahead scheduling. However, both processes do not imply a natural evolutionary process, and both are not mutually exclusive. For example, many cross-border integration processes started with long-term system planning collaboration and could lead to, for instance, the development of regional day-ahead markets and there are also examples of focusing on short term markets only.

5 Investment Patterns and Trends and Its Effect on Electricity Retail Sector

The investments in the east coast Australian electricity industry had been driven by policy and price signals since its inception. As discussed above, prior to the 1990s, the electricity industry was vertically integrated within the government-owned state electricity commissions with the generation, transmission, and distribution including retail supply. With the establishment of the NEM, the generation and retail supply were separated and transformed into competitive components. However, transmission and distribution retained its monopoly characteristics.

In terms of pricing, the period of 1955 to 1980 experienced a fall in real electricity prices due to economies of scales generated due to the construction of large thermal power stations utilizing low-cost domestic coal and gas. To cite an example, the price in New South Wales (NSW) fell from $375/MWh to $164/MWh (56% decrease). On the other hand, the decline in prices in Queensland was 43% from $350/MWh to $210/MWh. However, between 1982 and 1986, the prices increased by approximately 20% in real terms and 60% in nominal terms. This increase was caused by increased investments in the capacity by state-owned electricity commissions and the cost was passed on to the customers (Rai & Nelson, 2019).

The establishment of the NEM has resulted in decentralizing the operational and investment decision-making from central planners comprising governments and regulators to commercial entities. This process not only promoted competition, but also allocated risk to the parties that have the best information, expertise, and incentives to manage such risks. From 1998 to 2009, more than 6,000 MW of gas-fired energy generation (both intermediate and peaking) entered the NEM, while half of it amounted to approximately 3,600 MW entered in 2008–2009 alone in response to policy signals and drought-induced price spikes of 2007–2008. The policy signals were in the form of the Queensland Gas Scheme (QGS) and the NSW Greenhouse Gas Abatement Scheme. The QGS, which commenced in 2005, required generators to source 13% of their Queensland electricity from gas-fired generation. Generally, the process remained stable for the consumers due to quiescent wholesale prices and flat network prices. During this period, real residential prices increased 6% and 8% in NSW and Queensland, respectively (Rai & Nelson, 2019). The inflation-adjusted electricity retail price index is presented below in Table 8.8 with the base year of 2000.

The Australian Consumer and Competition Commission (2018) and Wood et al. (2018) have attributed the increase in consumer electricity prices

TABLE 8.8 Electricity retail price inflation adjusted index (2000–2019)

Year	Brisbane	Sydney	Melbourne	Adelaide	Hobart	Canberra	National
2000	100	100	100	100	100	100	100
2001	101.0	97.0	106.7	100.8	99.0	101.0	100.7
2002	100.5	96.8	106.6	100.0	100.3	98.3	100.1
2003	101.8	97.4	108.3	122.8	100.8	107.1	103.4
2004	101.9	103.0	104.3	116.7	101.6	105.8	103.4
2005	101.7	107.1	101.5	110.3	99.3	106.7	102.8
2006	102.5	109.1	98.1	109.0	99.7	106.7	102.1
2007	111.3	115.8	98.4	107.5	102.1	123.5	105.8
2008	114.2	120.2	109.3	114.8	114.2	125.1	111.0
2009	129.4	144.4	119.9	117.6	118.6	132.3	126.9
2010	142.2	153.5	139.7	120.8	122.3	131.8	138.7
2011	145.9	170.3	151.9	145.0	143.1	134.0	150.8
2012	162.7	196.3	183.6	178.2	157.5	157.0	175.3
2013	189.3	201.5	190.4	170.0	157.2	158.7	182.0
2014	190.8	182.7	175.1	168.5	134.5	143.2	170.1
2015	190.7	171.2	181.9	153.1	135.7	135.8	167.5
2016	193.9	186.0	175.9	168.8	138.6	142.4	173.2
2017	197.7	210.4	186.8	206.0	137.1	154.1	189.6
2018	175.4	204.7	202.8	197.9	136.2	164.6	189.4
2019	173.3	195.7	188.9	191.5	135.9	162.7	182.5

Source: ABS (2020)

TABLE 8.9 Change in average residential electricity customer prices in the NEM (cents per kWh)

	2007–2008	2017–2018
Wholesale	7.43	9.67
Network	8.8	12.2
Environmental	0.4	2.3
Retail costs	1.7	3.1
Retail margin	1.07	1.73
Total	19.4	29

Source: ACCC Inquiry into the National Electricity Market (August 2019 report)
Note: Based on effective unit charges paid by residential customers. Data is inflation adjusted, in 2017–2018 dollars, and excludes GST.

to the following. There was an increase in network expenditures, especially at the distribution level, as real prices increased by 85% between July 2008 and July 2018 (Table 8.9), of which the network component accounted for two-fifths of the increase, amounting to $60/MWh.

The composition of the residential bill in 2017–2018 is given in Table 8.10.

Uncertainty in emission reduction targets and trajectories with rising gas prices resulted in barriers to entry for gas-fired plants. The government emission reductions policies were aimed at zero-emission technologies such as Renewable Energy Targets (RET) and various other government schemes. This lack of technology neutrality for achieving emission reductions resulted in higher wholesale prices and precluded cheaper forms of emissions abatements. The production subsidies for zero-emission technology resulted in disorderly exit of plants as it did not provide an economic signal for efficient exit of plants, thereby raising the costs in the NEM to achieve lower carbon targets. Environmental costs accounted for one-eighth of the increase ($20/MWh) of the total 85% increase during July 2008 to July 2018.

Government policies and rising fuel costs led to gas-fired plants' lack of interest in participating in the NEM, despite high price signals in the form of higher wholesale prices. Simshauser (2014) argued that, between 2006 and 2015, the size of the regulatory asset base (RAB) increased from $47 billion to $82.5 billion. The RAB was also increased due to a tightening of network reliability and bushfire standards in NSW and Queensland without paying due consideration to the consumers' willingness to pay for increased investments in the networks. Nevertheless, these investments resulted in an increase in network revenues and prices. Due to the increase in RAB, the network prices increased by 85% (Rai & Nelson, 2019). Table 8.11 facilitates comparison of actual transmission and distribution against transmission forecast and distribution forecast.

TABLE 8.10 Composition of a residential electricity bill (cents per kWh) (real $ 2017–2018)

	Victoria	NSW	ACT	Queensland	South Australia	Tasmania	NEM
Wholesale	9.13	8.97	9.94	9.93	14	10	9.7
Network	11.53	12.36	7.18	12.53	13.2	10.4	12.2
Environmental	2.4	2.1	3.29	1.5	4.5	2.3	2.3
Retail costs	4	2.8		3.1	3.3		3.1
Retail margin	2.1	2.3		0.6	0.6		1.7
Retail costs and margin			3.26			2	
Retail electricity price	29.16	28.53	23.67	29.4	35.6	24.7	29

Source: ACCC (2018), Residential electricity price trends review, final report (December)

Note: Average residential customer prices excluding GST (real $2017–2018). Retail costs and margins are combined for the ACT and Tasmania due to data availability.

TABLE 8.11 Network investment – forecast and actual (2019 $ million)

	Transmission Forecast	Distribution Forecast	Transmission Actual	Distribution Actual
2006	802.5	3,996.2	913.5	4,521.0
2007	762.4	4,530.6	1,007.2	4,876.9
2008	1,661.0	4,793.3	1,715.2	4,872.2
2009	1,629.7	4,954.0	1,814.4	5,854.1
2010	1,799.6	6,461.0	1,477.7	6,293.9
2011	1,846.8	7,830.4	1,474.1	6,958.5
2012	1,749.9	8,022.1	1,642.1	7,081.6
2013	1,791.3	7,957.7	1,601.5	6,455.6
2014	1,667.5	7,897.4	1,268.1	5,417.4
2015	1,120.4	6,144.0	759.8	4,880.5
2016	1,310.2	4,806.4	755.9	3,655.2
2017	1,229.4	4,821.3	767.0	3,716.8
2018	721.2	4,507.6	757.9	4,073.5
2019	692.3	4,204.2		
2020	751.8	4,304.3		

Source: State of the energy market 2019 update

Note: Actual outcomes for relevant year on an end of year basis, 2019 dollars. Capital expenditures data are actual outcomes to 2018, and forecasts for 2019 and 2020. Networks report on a 1 July to 30 June basis, except in Victoria, where they report on calendar year basis for distribution and 1 April–31 March basis for transmission.

In addition to over-investments in RAB due to tightening of reliability and security standards, investments were also made based on over-estimated demands that contributed significantly in increased retail prices for the customers. Table 8.12 presents the grid demand consumption in the NEM.

The grid demand consumption increased a little after 2005, but it flattened in 2009–2010. Since this period, there has been a continuous decrease in grid consumption. The decrease in grid demand coincided with increased investments in the rooftop solar. Moreover, as the above table demonstrates, large investments were made during the period of the Global Financial Crisis (GFC) that increased financing costs. Consequently, higher rates of returns were required on the regulated assets, which were insured against any risk in the form of a guaranteed rate of return by the states. The dampening of the grid demand compounded the increase in network prices, implying that network price increased at a faster rate than the network revenue.

Market participants responded to the wholesale price signals in the market and insured their exposures to risk through hedging in the financial markets. Price signals pushed new investments into the forward contracts. However, distortions started building up when production subsidies designed solely for zero emissions were introduced to achieve policy goals of Renewable Energy

TABLE 8.12 Electricity consumption in the NEM (terawatts hours)

	Grid Demand	Rooftop Solar
2005–2006	202.756	0.006
2006–2007	207.036	0.009
2007–2008	208.620	0.016
2008–2009	210.481	0.043
2009–2010	209.844	0.155
2010–2011	207.492	0.690
2011–2012	203.380	1.728
2012–2013	198.208	2.958
2013–2014	193.610	4.019
2014–2015	193.968	5.052
2015–2016	197.603	6.100
2016–2017	196.496	6.302
2017–2018	196.160	7.301
2018–2019	195.689	9.168

Source: State of energy market 2019 update

Note: Grid consumption is native demand (including scheduled and semi-scheduled generation, and intermittent wind and large-scale solar generation)

Targets. This strategic policy choice of subsidizing certain forms of generation resulted in technology non-neutrality for the already-existing plants. The result was that cheaper forms of emissions abatement from reducing the emissions intensity of the existing plants were precluded, leading to an increase in wholesale prices (Simshauser & Akimov, 2019).

The Renewable Energy Targets provided incentives to maximize the output irrespective of the wholesale prices. However, the majority of the generators relied on wholesale price signals. Due to the federal and state governments' large-scale renewable energy target (LRET) and the small-scale renewable energy target (SRES) policies, the firm dispatchable plants retired before being replaced with the equivalent ones that constrained the supply of electricity in the system and drove up the prices (Table 8.13). Between June 2012 and June 2017, 4,255 MW of coal-fired plants exited the NEM and the weighted average notice period was 2.9 months with the highest notice period being 6.9 months (Rai & Nelson, 2019).

Helm (2014) has explained that in the event of an energy market crisis, inquiries were conducted that produced policy recommendations. Nevertheless, some of the policy recommendations were misguided because the market rarely provided an opportunity to scrutinize their unintended side effects. In this case, the misguided policy recommendation was to tighten network reliability standards. The combination of tightened reliability standard and load demand resulted in unprecedented investments in the network between 2005 and 2015. Due to the GFC and disruptive competition

TABLE 8.13 Generation withdrawals since 2012–2013

Year	Power Station	Region	Technology	Capacity (MW)	Status
Withdrawn				**4,174**	
2014–2015	Wallerawang C	NSW	Coal	1000	Retired
2014–2015	Morwell, Brix	Vic	Coal	190	Retired
2014–2015	Redbank	NSW	Coal	144	Retired
2014–2015	Callide A	NSW	Coal	30	Retired
2015–2016	Northern	SA	Coal	530	Retired
2015–2016	Playford B	SA	Coal	240	Retired
2015–2016	Collinsville	Qld	Coal	190	Retired
2015–2016	Anglesea	Vic	Coal	150	Retired
2015–2016	Barcaldine	Qld	CCGT	20	Downgraded
2016–2017	Hazelwood	Vic	Coal	1600	Retired
2016–2017	Mt Piper	NSW	Coal	80	Downgraded
Announced withdrawal				**2,755**	
2021	Torrens Island A	SA	Gas	480	Mothballing of units progressively between 2020 and 2022
2021	Mackay	Qld	OCGT	34	Retirement
2022	Daandine	Qld	CCGT	33	Retirement
2023	Liddell	NSW	Coal	2,000	Retirement
2050	Tamar Valley	Tas	CCGT	208	Retirement

Source: State of energy market 2019 update
Note: Data in August 2019. CCGT, combined cycle gas turbine; OCGT, open cycle gas turbine.

from distributed resources, the NEM experienced reduction in final electricity demand that resulted in what is termed a "network in decline" and was called a "utility death spiral" by Simshauser and Akimov (2019).

The role of interconnectors and transmission networks becomes important from an investment perspective when generators located in one region become marginal generators in another region for setting the price due to the integrated characteristics of the NEM. This happens when a mix of generators offer to supply electricity to the market at one time at a range of prices and the generator with the highest priced offer is required to meet the demand and thereby sets the dispatch price in the region and all other generators then receive the same dispatch price. For instance, a generator located in Victoria could set the price in NSW. The exit of Hazelwood in March 2016, which accounted for 5% of supply in the NEM, resulted in the withdrawal of low-cost supply from the market and was replaced by output from expensive generators (black coal, gas and hydro). Due to the closure of the Hazelwood, the black coal generators in NSW and Queensland increased output by 6% while gas-powered generators located in Victoria and South Australia increased output by 37%. Even though brown coal generators were in the market, they were running at full capacity and more expensive generators became the generators in setting the wholesale price in NEM (ACCC, 2018).

Table 8.14 shows the electricity retail market share among the participants of the NEM. The NEM is based upon competition among market participants to achieve efficient outcomes. Therefore, market concentration and factors contributing to increase the concentration basically compete against the fundamentals of the NEM market design. In fact, competition is crucial to achieve affordable prices in the energy-only market. Therefore, current and future investment in new generation capacity needs to be encouraged to achieve efficient outcomes in the form of affordable prices; however, it may not result in increasing the firm's market share as a result of new investment (ACCC, 2018).

A risk management tool is vertical integration where sellers and buyers vertically integrate to manage their risks. It has implications from an investment perspective. It stabilizes the exposure for both generator and retailer, because any spike faced by the retailer resulting in loss of revenue is offset by the generator, which earns additional revenue due to that spike. The degree to which the retailer or generator manages to mitigate their risk depends upon the size of their load or capacity to generate, respectively. The NEM has observed significant vertical integration between retailers and generators (Table 8.15). Other stakeholders have raised concerns about the effect of this vertical integration on hedging markets because, as discussed earlier, it affected the liquidity of contracts in the markets.

The Retailer Reliability Obligation (RRO) was introduced into the National Electricity Law (NEL) and National Electricity Rules with effect from 1 July 2019 when the National Electricity (South Australia) (Retailer Reliability Obligation) Amendment Act 2019 (SA) came into force (AEMO, 2019). Under the Retailer Reliability Obligation, the COAG tasked AEMO to forecast future energy demand that would provide a signal to the market participants of the expected shortfalls in future electricity supplies. As a result, electricity retailers would be making contracts to shield customers from sudden spikes in wholesale electricity prices. Retailers are required to maintain sufficient contracts to cater to electricity shortage as forecasted by AEMO. If a retailer does not hold sufficient contracts, the retailer could be liable to cover the cost of emergency actions undertaken by AEMO acting as a "Procurer of Last Resort" with potential penalties of up to $100 million (Michael, 2019).

The reliability requirement facilitates investment in dispatchable capacity by building on existing spot and financial market arrangements in the

TABLE 8.14 Electricity retail market share (number of small customers)

	Queensland	NSW	Victoria	South Australia	Tasmania	ACT	Total
Origin	565,645	1,127,283	527,097	225,874		25,780	2,471,679
AGL	355,853	806,263	599,397	334,288			2,095,801
EnergyAustralia	119,150	1,015,010	498,410	68,375		8,453	1,709,398
Snowy Hydro	47,991	234,386	422,733	44,800		40	749,950
Ergon	705,199						705,199
Simply	14,758	34,053	260,582	87,387		42	396,822
Aurora					280,122		280,122
Alinta	190,010	96,148	91,712	58,170			436,040
ActewAGL		28,527				154,580	183,107
Amaysim	39,391	53,848	40,674	6,500			140,413
Momentum	19	21,422	108,958	6,582			136,981
M2 Energy	10,896	32,802	48,712	5,366			97,776
Powershop	9,325	32,807	61,161	241			103,534
Other	158,277	102,274	140,670	31,458	1,098	488	434,265
Total	2,216,514	3,584,823	2,800,106	869,041	281,220	189,383	9,941,087

Source: State of energy market 2019 update

Note: Includes residential and small business customers. All data at June 2019, except Victoria (June 2018).

TABLE 8.15 Vertical integration in NEM jurisdictions

Queensland (%)

	Electricity Generation	*Electricity Retail*
Origin	2.46	29.40
EnergyAustralia	0.21	5.38
AGL	0.22	16.69
Engie (Simply)	0.00	0.67
Snowy Hydro (Red/Lumo)	0.00	2.17
Hydro Tas (Momentum)	0.00	0.00
Alinta	0.76	8.57
Other private	20.61	5.30
Other state-owned	75.74	31.82

NSW & ACT (%)

Origin	28.59	30.71
EnergyAustralia	14.30	27.12
AGL	39.37	21.94
Engie (Simply)	0.21	0.90
Snowy Hydro (Red/Lumo)	2.67	6.21
Hydro Tas (Momentum)	0.00	0.57
Alinta	0.00	2.55
Other private	14.16	5.15
Other state-owned	0.70	4.85

Victoria (%)

Origin	2.81	18.82
EnergyAustralia	25.78	17.80
AGL	42.27	22.95
Engie (Simply)	0.00	9.31
Snowy Hydro (Red/Lumo)	3.61	15.10
Hydro Tas (Momentum)	0.44	3.89
Alinta	20.74	3.28
Other private	3.89	8.86
Other state-owned	0.44	-

South Australia (%)

Origin	22.19	25.99
EnergyAustralia	1.82	7.87
AGL	31.77	41.28
Engie (Simply)	27.00	10.06
Snowy Hydro (Red/Lumo)	0.39	5.16
Hydro Tas (Momentum)	1.65	0.76
Alinta	0.00	6.69
Other private	5.60	2.20
Other state-owned	9.59	-

(*Continued*)

TABLE 8.15 (Continued)

	Electricity Generation	Electricity Retail
Tasmania (%)		
Origin	0	0
EnergyAustralia	0	0
AGL	0	0
Engie (Simply)	0	0
Snowy Hydro (Red/Lumo)	0	0
Hydro Tas (Momentum)	100	0
Alinta	0	0
Other private	0	0.39
Other state-owned	0	99.61

Source: State of energy market 2019 update
Note: Electricity generation market shares are based on generation capacity owned or controlled in January 2018. Retail market shares are based on number of small customers in June 2018, except Victoria (June 2017).

electricity market (Table 8.16). It incentivizes the retailers on behalf of customers to ensure the reliability of the power system through their contracting and investment in resources.

It is worth noting that Australia's electricity system of the future will be composed of many small and geographically dispersed renewable wind- and solar-powered generators. These generators historically do not have large amounts of transmission capacity. Hence, there will be a need for sufficient and appropriate transmission infrastructure for better coordination of generation and transmission investment decisions. This would facilitate a smooth transition and would minimize costs for end users of power. The transmission access reforms are being contemplated and the proposed access model envisages Locational Marginal Pricing (LMP) and financial transmission rights (FTRs). Under the LMP, large-scale generators and storage would receive a spot price that would vary with their locations. Retailers, and as a result customers, would continue to pay the regional reference price, which would promote contract market liquidity. Under LMP, electricity supply (generation) is priced drawing on local supply and demand conditions. Under FTR, participants would be able to purchase FTRs and pay out on the differences in wholesale market prices that arise due to congestion and losses. In order to ensure a smooth transition, some FTRs would be allocated ("grandfathered") for free. This will eliminate the sudden changes that would occur in the financial market and will provide time for market adjustment. Thus, the existing transmission congestion and loss-embedded risks for the market participants can be managed in the best possible way. Consequently, more revenue certainty and the confidence to invest will be firmly established (AEMC, 2020).

TABLE 8.16 Committed investment projects in the NEM

Developer	Power Station	Technology	Capacity (MW)	Planned Commissioning
Queensland			286	
MSF Sugar	Tableland Mill	Bagasse (expansion)	24	2019–2020
Risen Energy	Yarranlea	Solar	103	2019–2020
Windlab/Eurus	Kennedy Energy Park	Solar	15	2019–2020
Windlab/Eurus	Kennedy Energy Park	Battery	2	2019–2020
Windlab/Eurus	Kennedy Energy Park	Wind	43	2019–2020
Maryrorough solar	Maryrorough	Solar	35	2019–2020
University of Queensland	Warwick	Solar	64	2019–2020
NSW			1,009	
John Laing/ Maoneng Group	Sunraysia	Solar	229	2019–2020
Edify Energy; Octopus Investments	Darlington Point	Solar	275	2019–2020
Innogy	Limondale – Plant 2	Solar	29	2019–2020
Elliott Green Power	Nevertire	Solar	105	2019–2020
Spark Infrastructure	Bomen	Solar	121	2019–2020
Innogy	Limondale – Plant 1	Solar	220	2019–2020
TEC-C Investments	Molong	Solar	30	2019–2020
Commonwealth Government	Snowy 2.0	Pumped Hydro	2,040	2024–2025
Victoria			1,879	
Total Eren	Kiamal – Stage 1	Solar	200	2019–2020
Enel Green Power	Cohuna	Solar	31	2019–2020
Flow Power	Yatpool	Solar	94	2019–2020
Neoen	Bulgana Green Power Hub – BESS	Battery	20	2019–2020
Neoen	Bulgana Green Power Hub	Wind	204	2019–2020

(*Continued*)

TABLE 8.16 (Continued)

Developer	Power Station	Technology	Capacity (MW)	Planned Commissioning
Northleaf 40%; InfraRed Capital Partners 40%; Macquarie 20%	Lal Lal – Elaine end	Wind	84	2019–2020
Goldwind	Stockyard Hill	Wind	532	2019–2020
John Laing Group	Cherry Tree	Wind	58	2019–2020
Goldwind	Moorabool	Wind	320	2019–2020
Tilt Renewables	Dundonnell	Wind	336	2020–2021
South Australia			306	
AGL Energy	Barker Inlet	Gas	210	2019–2020
Nexif Energy	Lincoln Gap	Battery	10	2019–2020
Nexif Energy	Lincoln Gapi–stage 2	Wind	86	2019–2020
Tasmania			266	
Granville Harbour Operations	Granville Harbour	Wind	112	2019–2020
Wild Cattle Hill	Wild Cattle Hill	Wind	154	2019–2020

Source: State of energy market 2019 update
Note: Data at 1 October 2019.

6 Conclusions

6.1 Lessons Concerning Market Integration for the NER

From the experience of the evolution of NEM, the following lessons can be drawn for the NER:

1. There should be clarity of purpose as the process of integration entails economic and policy implications, commercial and financial impacts, operational and organization changes, which are all required to be brought into alignment.
2. The experience of NEM has also demonstrated the importance of political drive and commitment, as it required not only leadership time and energy but also financial incentives for the participating organizations to make the NEM process a viable option.
3. Appropriate governance structure was established among the participating jurisdictions to ensure coordination of policy, technical design, and implementation.

4. A bottom-up approach was adopted by implementing reforms at the state level before moving to a full national electricity market. This approach helped in building up confidence among the participants as state-level learning experiences helped in improving the market technical designs.

5. Transparency and involvement of all stakeholders at the broader scale, along with the pace of reform through an open dialogue process, help the reform process to become manageable and realistic. The process created ownership among all stakeholders as all stakeholders were able to devote resources to it.

6. The process of competition was followed in two dimensions, namely vertical and horizontal. In the vertical dimension, competition was ensured in the restructuring of electricity monopolies by separating generation and retail from the natural monopolies' segments of transmission and distribution. In the horizontal dimension, competition was introduced in the generation and retail segments to reap the benefits of competition.

7. There was a tradeoff between introducing competition in the electricity sector and generation revenue from the privatization proceeds. By introducing competition in the sector, the financial impact was endured to ensure the sustainability of the reform process.

8. Competition laws were amended, and measures were taken to ensure liquidity in the financial markets.

9. The share of renewables is increasing in the generation mix due to increased investments over the past few years in NEM. This has brought forward the increased focus on the objectives of ensuring reliability and security of the power systems. NEM has kept pace with these developments and established Energy Security Board was established in 2017.

10. In the operation of electricity system, it is important to establish a link between the physical aspect of electricity generation and financial markets as this enables the participants to change their electricity generation patterns according to financial incentives. This brings in market-driven efficiency dynamics in the electricity generation process.

11. The real-time dispatch of electricity through market design, appropriate governance structures and competition under the integration process has ensured the least cost production of electricity.

12. Regional planning plays an important part in directing the flow of investments to ensure reliability and security of the power sector.

13. Investments in the transmission and distribution segments should follow an objective criterion of regulatory investment tests to ensure net market benefits. This framework would enable the investments to be directed towards efficient utilization of resources.

6.2 Lessons Concerning Investment Pattern for the NER

The experience of the NEM has highlighted a number of important challenges from an investment perspective that need to be carefully considered in the design of the NER.

1. The introduction of solar photovoltaic (PV) has not only reduced the energy grid demand, but also has adversely affected the retail tariffs that have further contributed to reductions in demand. This aspect needs to be carefully considered in the design of the NER.
2. Production subsidies given to renewables in order to achieve emission targets have resulted in a disorderly exit of the firm dispatchable plants before they could be replaced with equivalent ones. This has constrained the supply of electricity in the system and has driven up the prices. A technology-neutral approach could be an important area of consideration in the development of the NER.
3. The NEM has followed the path of strict reliability standards that has resulted in substantial investments in the transmission and distribution networks, even during the period of stagnating or falling demand. As a result, the network weighted cost of capital has significantly affected rise of tariffs faced by the customers. The NER needs to consider the tradeoff between the strict reliability standards and its effects on the retail prices faced by the customers.
4. Financial markets need to function efficiently to buffer the risks faced by both buyers (retailers) and sellers (generators). The introduction of renewable energy targets, Feed-in Tariff (FiT) schemes, and other green policies has distorted the price signals in the NEM due to tightening of supplies and reducing the liquidity of contracts in the markets.
5. There has been vertical integration in the NEM with both retailers and generators trying to stabilize their exposure to risk and mitigate any adverse consequences due to price spikes. However, the related concerns are about the concentration of incumbents in terms of market share to influence the prices and the reduction of contract liquidity in the market. In the design of the NER, clear rules could be delineated in this regard right from the start of the market design.
6. Due to the integrated characteristic of the NEM, a generator located in one region could become the marginal generator in another region and set the price of the market. Due to interconnectedness, electricity will flow from one region to another region, which sometimes strains the interconnectors given the capacity of the transmission systems, and could also lead to an expensive source of electricity generation. While considering investments in the NER, these aspects also need to be considered in the market design.

7. While smoothening and facilitating the path for future investments to ensure the reliability and security of the electricity system, forecasts and other related information are provided to potential investors by the AEMO. The NER also needs to develop a mechanism to provide information to the potential investors so that there should not be information asymmetries in their investment decision-making.
8. The NEM has been contemplating the locational marginal pricing and financial transmission rights due to the recent developments in the form of relatively small and geographically dispersed renewable generators, connecting to windy or sunny parts of the network, which historically have not needed large amounts of transmission capacity. So new investments would be required in the transmission and distribution sectors, encouraging the move toward regional reference pricing that is likely to promote contract market liquidity and also reduce congestion and loss-related risks. The NER also needs to consider these options at the design state of the market design.

Note

1 These working groups (WG) with broad representation of all stakeholders involved in the electricity value chain composed of, for example, Market Trading WG, Transmission Pricing WG, and National Electricity Code WG.

References

ACCC. (2018, June). *Restoring electricity affordability and Australia's competitive advantage, retail electricity pricing inquiry – Final report.* Australian Competition and Consumer Commission, Canberra.

Alan, R., & Nelson, T. (2019). Ausetralia's national electricity market after twenty years. *Australian Economic Review, 52*(2), 165–182.

Australia Energy Market Operator. (2019). 2019 electricity statement of opportunities: A report for the national electricity market. Retrieved August 1, 2020, from https://aemo.com.au/-/media/files/electricity/nem/planning_and _forecasting/nem_esoo/2019/2019-electricity-statement-of-opportunities.pdf?la =en&hash=7FE871D75A9C619AB66FA671477551B2.

Australian Consumer and Competition Commission. (2018), Restoring electricity affordability and Australia's competitive advantage. ISBN 978 1 920702 34 2. https://www.accc.gov.au/system/files/RetailElectricity+Pricing+Inquiry%E2 %80%94Final+Report+June+2018_0.pdf

Australian Energy Market Commission (AEMC). (2020). Retrieved May 20, 2020, from http://www.aemc.gov.au/.

Australian Energy Market Commission. (2020). *Transmission access reform update paper.* Retrieved August 1, 2020, from https://www.aemc.gov.au/sites/default /files/2020-03/March%20update%20paper%20-%20transmission%20access %20reform_0.pdf.

Australian Energy Regulator (AER). (2018). *State of the Energy Market*, Chapter1, Retrieved 20 May, 2020, from https://www.aer.gov.au/system/files/State%20of

%20the%20Energy%20Market%202018%20-%20Chapter%201%20A3
%20spread_1.pdf.

Australian Energy Regulator (AER). (2019). *State of the energy market 2019*. Retrieved May 20, 2020, from https://www.aer.gov.au/publications/state-of-the-energy-market-reports/state-of-the-energy-market-%E2%80%93-data-update-november-2019.

Fitzgerald, P., & Dreyfus, S. (1995). *The price of power*, consultancy report to the churches, Melbourne: Brotherhood of St Laurence.

Helm, D. (2014). Electricity and energy prices, Energy Futures Network paper. Retrieved from http://www.dieterhelm.co.uk/assets/secure/documents/Electricity-and-energy-prices.pdf.

Higgs, H., & Worthington, A. C. (2016). The impact of design rules on wholesale electricity prices in the Australian national electricity market (NEM). In *Electricity markets: Impact assessment, developments and emerging trends*. Griffith Research Online, Queensland, https://www.novapublishers.com/catalog/product_info.php?products_id=59038.

International Energy Agency (IEA). (2001). *iEnergy policies of IEA countries, Australia 2001 Review*. Paris: Organization for Economic Cooperation and Development (OECD).

International Energy Agency (IEA). (2019). *Integrating power systems across borders*. Retrieved May 22, 2020, from https://www.iea.org/reports/integrating-power-systems-across-borders.

KPMG. (2013). *National electricity market: A case study in successful microeconomic reform*. Retrieved May 23, 2020, from https://www.aemc.gov.au/sites/default/files/content/The-National-Electricity-Market-A-case-study-in-microeconomic-reform.PDF.

Michael, M. (2019, June). iAustralia gets half a NEG as COAG signs off on reliability obligation. *Renew Economy*. Retrieved August 1, 2020, from https://reneweconomy.com.au/australia-gets-half-a-neg-as-coag-signs-off-on-reliability-obligation-45613/.

Pierce, J. (2013). *National electricity market: A case study in successful microeconomic reform*. Retrieved May 23, 2020, from https://www.aemc.gov.au/sites/default/files/content/The-National-Electricity-Market-A-case-study-in-microeconomic-reform.PDF.

Productivity Commission (PC). (1991). *Energy generation and distribution*. Industry Commission inquiry report 1991. Retrieved May 2020, from https://www.pc.gov.au/inquiries/completed/energy-generation.

Siemon, D. (1995). *The restructuring and sale of Victoria's electricity industry: Is it worth it?* Melbourne: Brotherhood of St Laurence. Retrieved May 23, 2020, from http://library.bsl.org.au/jspui/bitstream/1/1658/1/The%20restructuring%20and%20sale%20of%20Victoria's%20electricity%20industry.pdf.

Simshauser, P.i (2014). From first place to last: The national electricity market's policy- induced energy market death spiral. *Australian Economic Review*, 47(4), 540–562.

Simshauser, P., & Akimov, A. (2019). Regulated electricity networks, investment mistakes in retrospect and stranded assets under uncertainty. *Energy Economics*, 81, 117–133.

Skinner, B. (2018). *Happy 20th birthday national electricity market*. Australian Energy Council. Retrieved May 23, 2020, from https://www.energycouncil.com. au/analysis/happy-20th-birthday-national-electricity-market/.

State of Energy Market. (2019). https://www.aer.gov.au/publications/state-of-the -energy-market-reports/state-of-the-energy-market-%E2%80%93-data-update -november-2019

Stockdale, A. (1994). '*Enterprise Victoria: recapturing our low-cost energy advantage*', unpublished notes of presentation, Melbourne.

Wood, T., Blowers, D., & Percival, L. (2018, June). *Mostly working: Australia's wholesale electricity market*. Grattan Institute.

INDEX

Note: Page numbers in *italic* refer to figures and **bold** refers to tables

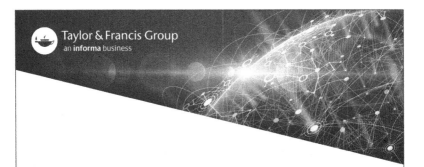

Printed in the United States
by Baker & Taylor Publisher Services

Printed in the United States
by Baker & Taylor Publisher Services